Activity Boo

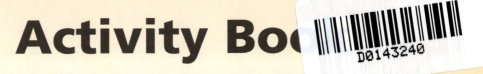

Michael Tammaro
University of Rhode Island

PHYSICS MATTERS

James Trefil
George Mason University

Robert M. Hazen
George Mason University

WILEY

JOHN WILEY & SONS

Cover Photo: ©Lawrence Englesberg/Image State

To order books or for customer service call 1-800-CALL-WILEY (225-5945).

ISBN 0-471-42898-1

Printed in the United States of America

10 9 8 7 6 5 4 3 2 1

Printed and bound by Courier-Kendallville, Inc.

Contents

Activity 1.1

Proportionality

In this activity you will explore the concept of proportionality by measuring the circumference and diameter of four round objects, and graphing your results. The diameter of a circle is the distance across the circle, and the circumference is the distance around the circle.

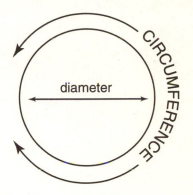

COLLECT MATERIALS

You will need the following materials for this activity: a piece of string, a ruler, and four round objects.

TAKE MEASUREMENTS

Measure the diameter and circumference of four round objects and record the data in the table below. Wrap the string around the outside of the objects to find the circumference. Be sure to use the same units for each of your measurements.

OBJECT	CIRCUMFERENCE	DIAMETER

What units (inches, centimeters, etc.) did you use for your measurements?

_____

What trends do you notice in your data table?

_____

_____

_____

PLOT YOUR DATA

Note the range of the values for circumference in the table. Mark the vertical axis with an appropriate scale. Do the same for the horizontal axis. Use the blank graph below to plot your data.

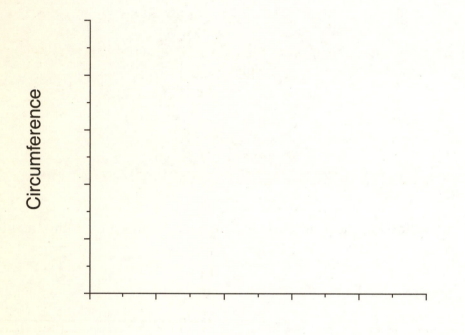

FIND THE SLOPE OF THE GRAPH

Use a ruler to draw a straight line that best fits the data. Find the *slope* of the graph. The slope of a graph is the vertical rise divided by the horizontal run.

What is the slope?

_____

If your data is well-fitted by a straight line, then we can say that the circumference of a circle is directly proportional to the diameter. Using the slope of your graph, write an equation that relates the circumference of a circle to its diameter:

Activity 1.2

A Scientific Experiment

COLLECT MATERIALS

Find a coin, a smooth inclined plane (a hard-covered book propped up on one end will work just fine), and a ruler. Place the inclined plane at the edge of a desk or table. Be sure the incline is high enough so that the coin will slide off when released from rest.

We want to determine how the range (R) of the coin's flight depends on the height (h) at which it is released from the bottom of the inclined plane.

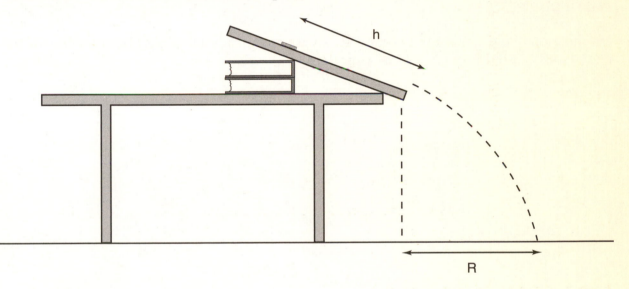

MAKE A HYPOTHESIS

Come up with a hypothesis for how the range will depend on the height. (Will the range be proportional to the height? Will the range be independent of the height?) Write your hypothesis below.

_____

_____

_____

_____

DO YOUR MEASUREMENTS

Release the coin, starting from rest, at five different values of h and measure the range. Do a few test trials first to get a feel for the measurements you will be making. If it is difficult to see exactly where the coin lands, you can rub some pencil lead on the coin and have it land on a piece of white paper. In this way, it will leave a clear mark when it lands.

What units will you be using for the measurements? (cm, in, feet, etc.)

_____

Record your data for the experiment in the table below.

h					
R					

PLOT YOUR RESULTS

On the graph below, mark the vertical axis with an appropriate scale. Do the same for the horizontal axis. Use this graph to plot your data.

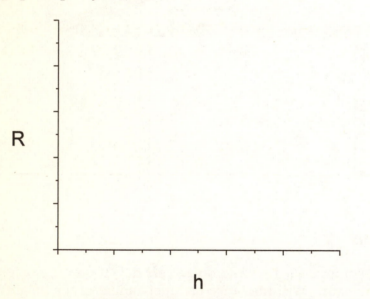

Do your results support your hypothesis? Why or why not?

_____

_____

_____

_____

_____

There was friction between the coin and the inclined plane. Do you think that if there was no friction, the graph would look different? Explain your reasoning.

_____

_____

_____

_____

Suppose there was no friction and the inclined plane was very long, so that the coin was moving very fast when it reached the bottom of the plane. What effect would this have on your results?

_____

_____

_____

_____

Make a rough sketch of what you think the graph would look like for a long, frictionless inclined plane.

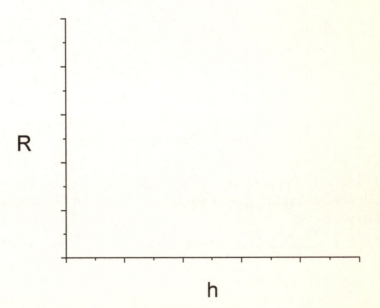

Activity 2.1

Graphic Language

Mathematics is the language of science, and mathematical relationships can usually be expressed graphically. Most of the time, we are interested in how one quantity depends on another. In this case, a simple graph with horizontal and vertical axes will do the trick.

In this activity, you will be given two quantities and will be asked to make an *approximate* graph of how one quantity changes with the other. Afterwards you will state what kind of a mathematical relationship would be appropriate to represent your graph; direct, inverse, or power law. Review chapter two as necessary.

The following is an example to help you get started.

EXAMPLE: The more gas you have in your gas tank, the further you can drive. Make a graph of the distance you can travel versus the amount of gas in your tank.

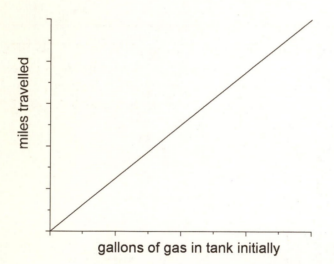

This is a *direct* relationship, so the graph is a straight line. We say that the number of miles traveled is *proportional* to the amount of gas in the tank. If you double the amount of gas in your tank, you can travel twice as far.

1. Suppose the jackpot of the state lottery is up to $100,000,000 and you want to increase your chances of winning by buying multiple tickets. Make a rough graph of the probability of winning the lottery versus the number of tickets you buy.

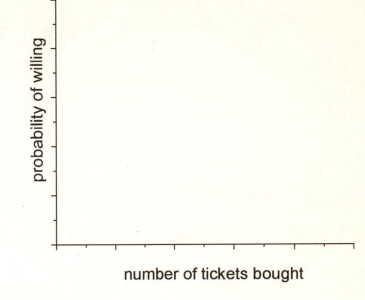

What kind of relationship is this?

_____

2. You have been chosen by your classmates to represent the school in a pie eating contest. Your goal is to eat as many pieces of pie as possible. Your strategy is to eat as fast as possible until you cannot eat any more. Make a graph of the number of pieces of pie you eat per minute versus the time since you started eating.

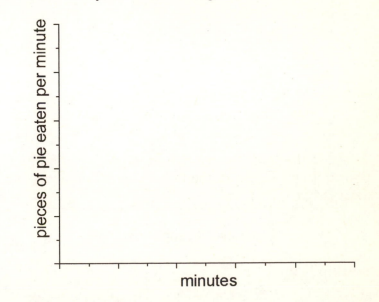

What kind of relationship is this?

_____

3. Oddly enough, your specialty is painting square walls. In order to make preparing for a job easier, you would like to represent, graphically, the amount of paint you must bring to the job versus the width of the wall. Make this graph.

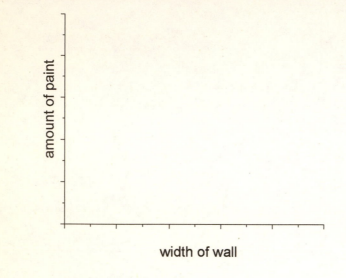

What kind of relationship is this?

_____

4. You're throwing yourself a big birthday party. In order to make the party as big as possible, you tell each person you invite that they may bring five friends. Assuming everyone that is invited shows up, graph the number of people at the party versus the number of people you invite.

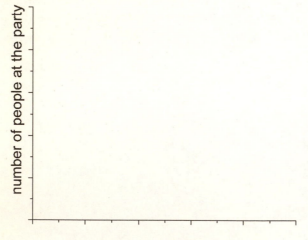

What kind of relationship is this?

_____

Activity 2.2

Scalars and Vectors

Any quantity that can be expressed as a single number is called a *scalar*. The amount of money in your pocket is a scalar quantity because it can be represented by a single number.

A *vector* is a quantity that has a direction. The location of the closest gas station is a vector quantity. For example, it may be 1.5 miles north. To uniquely specify the location, you need both the quantity (1.5 miles) and the direction (north).

Which of the following are vectors and which are scalars? (circle one)

the correct time	vector	scalar
the force exerted on a baseball by a bat	vector	scalar
the top speed of your car	vector	scalar
the location of Mt. Everest	vector	scalar
the temperature of boiling water	vector	scalar

Vectors can be represented by arrows. The length of the arrow is proportional to the magnitude (number). In this exercise, draw an arrow that best represents the velocity. Be sure that the lengths of the vectors are approximately correct with respect to each other.

a. 20 miles/hour, north

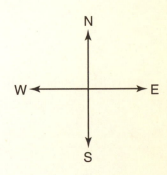

b. 30 miles/hour, southeast

c.   10 miles/hour, west

It is very common that two vectors must be added together. For example, two forces on an object may be represented as a single force that has the same effect as the two acting in combination. To add two vectors, connect the head of the first vector to the tail of the second. The sum of the two vectors is an arrow that connects the tail of the first vector to the head of the second.

In this exercise, you will be given two forces. Draw arrows to represent the two forces, and then draw an arrow that represents the sum of the two forces.

a.   10lbs to the right, 20lbs to the left

b.   10lbs up, 20lbs to the right

c.   30lbs to the right, 5lbs down

For the following vectors, draw two vectors (one horizontal and one vertical) whose sum would equal the vector shown.

a.

b.

c.

Activity 3.1

Paper-henge

This activity requires a sunny day, and a room with sun exposure for part of the day. First, cut out a triangular piece of paper and tape it to the middle of the window, so that the tip of the triangle casts a clear shadow somewhere in the room.

Tape an 8"x11" piece of paper in a location of the room so that the tip of the shadow is located in the center of the paper. Try to place the paper about one to two meters from the window. Make a small mark at the tip of the shadow and write the exact time next to it.

For the next hour, or until the shadow is off the paper, mark the location of the tip of the shadow every 5 minutes, labeling the time next to the mark. Staple the paper to this sheet before handing it in.

How far from the window was the piece of paper?

_____

What would you expect if you left the paper in place, and traced the shadow again tomorrow at the same times? If you expect a different trace, draw a dotted line on the paper to show what you expect, and label it 'tomorrow.'

_____

_____

_____

_____

_____

What would you expect if you left the paper in place, and traced the shadow again two months from now? If you expect a different trace, draw a dotted line on the paper to show what you expect, and label it 'two months from now.'

_____

_____

_____

_____

_____

In what way is your 'dorm-henge' different from Stonehenge, as discussed in chapter 3?

_____

_____

_____

_____

_____

Activity 3.2

Kepler's Second Law

The following diagram represents a satellite executing a circular orbit around a planet.

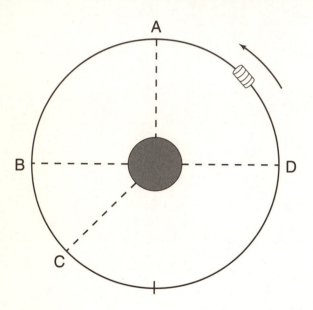

If the satellite takes 12 hours to go from A to B, how long would you expect it to take to go from C to D? Explain your reasoning.

_____

_____

_____

How long would you expect it to take to go from B to C? Explain your reasoning.

_____

_____

_____

Name ______________________ Date ____________ Class____________

The following diagram represents a satellite executing an elliptical orbit around a planet.

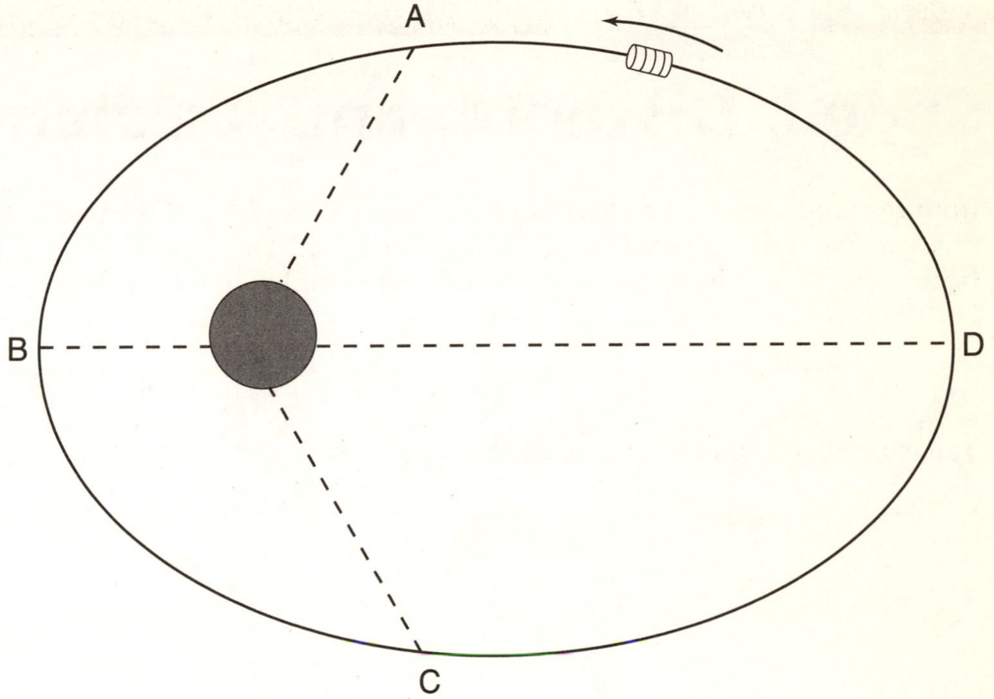

If the satellite takes 10 hours to go from A to B, how long would you expect it to take to go from C to D? Explain your reasoning.

____

____

____

How long would you expect it to take to go from B to C? Explain your reasoning.

____

____

____

Activity 3.3

Speed, Distance, and Acceleration

A car starts from rest and accelerates at a rate of 2 m/s per second for 10 seconds. For the next 10 seconds, the car maintains a constant speed. For the next 20 seconds, the car decelerates (accelerates backwards) at a rate of 1 m/s per second, until coming to a complete stop. The total elapsed time of the trip is 40 seconds.

When you answer the following questions, be sure to show your calculations.

How far did the car travel in the first 10 seconds?

How fast was the car traveling after the first 10 seconds?

How far did the car travel between 10 and 20 seconds?

How far did the car travel during the last 20 seconds?

Make a graph of the acceleration versus time for the entire trip.

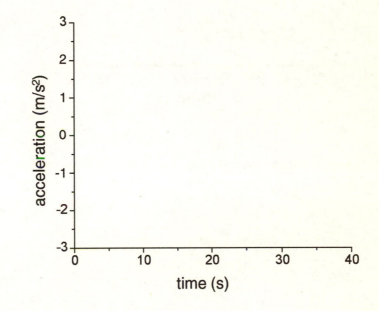

Make a graph of the speed of the car versus time.

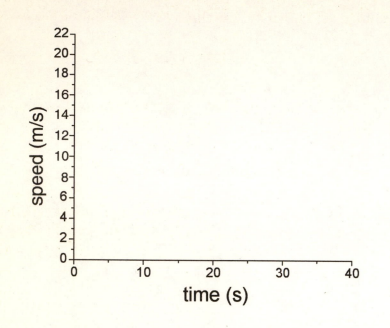

Make a graph of the distance traveled versus time.

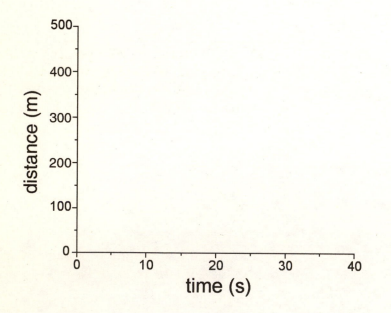

Activity 3.4

Free Fall

The acceleration of gravity is equal to 9.8 m/s², or 32 ft/s². In this activity you will investigate freely falling bodies. You may need to use the following equations:

$v = at$ and $D = \frac{1}{2}at^2$ (which can be written as $t = \sqrt{\frac{2D}{a}}$)

Suppose you drop a coin from rest, at eye level. Use the appropriate equation above and calculate the time it should take for the coin to fall to the ground. The distance from the ground to eye level is approximately your height.

Observe a clock or watch that displays seconds. Count the seconds in your head to get a feel for how long one second is. Stand up and hold a coin at eye level and drop it. How long would you say it took to fall to the ground? Does the time you calculated above seem reasonable?

_____

_____

_____

_____

Get a loosely crumpled piece of tissue or bathroom paper and drop it from eye level. How long would you say it took to fall to the ground? Did it take more or less time than the coin? (If you can't tell, then drop them simultaneously.)

_____

_____

According to the equation $t = \sqrt{\dfrac{2D}{a}}$, all things dropped from the same height would hit the ground at the same time. Would the coin and the tissue paper hit the ground at the same time when dropped simultaneously from eye level? If not, why not?

_____

_____

_____

_____

_____

Try dropping the coin and the tissue paper simultaneously from a height of 6 inches to a table-top. Can you tell which one hits first? It should be difficult to tell. Why, when dropped from a smaller distance, is the time almost the same?

_____

_____

_____

_____

_____

Activity 4.1

Newton's First Law

Close your eyes and try to feel all of the forces acting on you. Write them down in the space below.

According to Newton's First Law: A moving object will continue moving in a straight line at a constant speed, and a stationary object will remain at rest, unless acted upon by an unbalanced force.

Assuming that you are stationary, Newton's first law says that you must be acted on by a *balanced* force. In the context of the forces you listed above, what does it mean that you are acted upon by a *balanced* force?

_____

_____

_____

_____

_____

Drop something and watch it fall. Does it move in a straight line? _____ Does it move with constant speed? _____

What does Newton's First Law tell you about the force(s) acting on the object as it falls?

_____

_____

_____

_____

Imagine being in a car traveling along a straight road at constant speed. List all of the forces acting on the car.

If the car is moving in a straight line and at constant speed, the forces must be balanced. What does that tell you about the forces that you listed above?

_____

_____

_____

_____

Now suppose the driver of the car applies the brakes, and the car begins to slow down. List all of the forces acting on the car.

Are the forces balanced, or unbalanced now? What does that tell you about the forces acting on the car?

_____

_____

_____

_____

Activity 4.2

Newton's Second Law

A force of 4.45 newtons is equal to about one pound of force. A mass of one kilogram has a weight of 9.8 newtons on the surface of the Earth.

What is your weight in newtons?

What is your mass in kilograms?

If you travel to the moon, would your weight, your mass, or both change?

_____

_____

Suppose there is a 10kg object on a frozen pond. One rope is pulling it to the right, and one to the left.

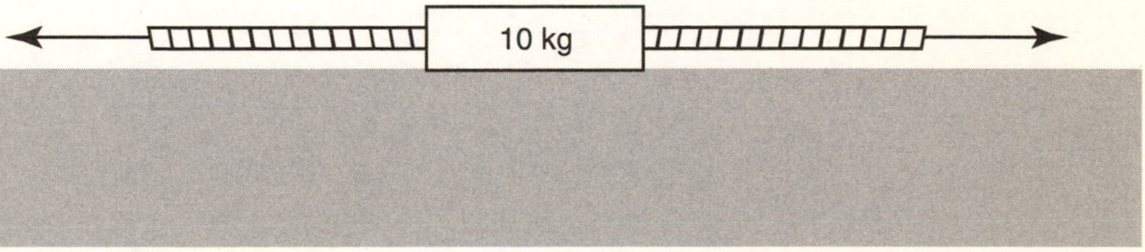

Fill in the missing information in the table.

Force pulling right	Force pulling left	Acceleration	Mass
10N	10N		10kg
	3N	10m/s²	5kg
5N		1m/s²	1kg
100N	50N	2m/s²	

What is the difference between speed and velocity?

_____

_____

What does it mean for an object to be accelerating to the right?

_____

_____

If an object is accelerating, is it necessarily moving? Explain.

_____

_____

_____

If an object is moving, is there necessarily a force on that object? Explain.

_____

_____

_____

_____

If an object has no acceleration, can there be forces acting on that object? Explain.

_____

_____

_____

_____

Activity 4.3

Newton's Third Law

Newton's third law states: *For every action (force) there is an equal and opposite reaction (force).* Thus, all forces come in pairs. There is an even number of forces in the universe. In this activity you will be given a variety of situations where forces are acting. When you are asked for a reaction force, state the force, and its direction.

Consider the forces on your textbook as it sits on a table. Assume that your textbook weighs 12 Newtons.

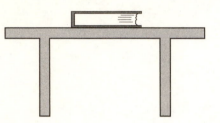

The table is pushing up on the book with a force of 12 Newtons. What is the reaction force?

_____

_____

Gravity is pulling down on the book with a force of 12 Newtons. What is the reaction force?

_____

_____

Let's consider a different situation. A man is having a tug-of-war contest with a young boy. The mass of the rope is negligible. In an attempt to pull the boy over the line, the man pulls suddenly with a force of 70 pounds.

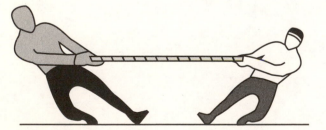

The man is pulling on the rope with a force of 70 pounds. What is the reaction force?

_____

_____

Provided the rope is not slipping in the boy's hand, what is the magnitude of the force of the rope pulling on the boy?

_____

You are riding your bicycle along a straight stretch of road. You are accelerating forward, so your speed is increasing.

You push down on the pedal with a force of 600 Newtons. What is the reaction force?

_____

_____

The road is pushing forward on your rear tire with a force of 500 Newtons. What is the reaction force?

_____

_____

Air resistance is pushing back on you with a force of 300 Newtons. What is the reaction force?

_____

_____

A baseball player swings at the ball and hits a home run.

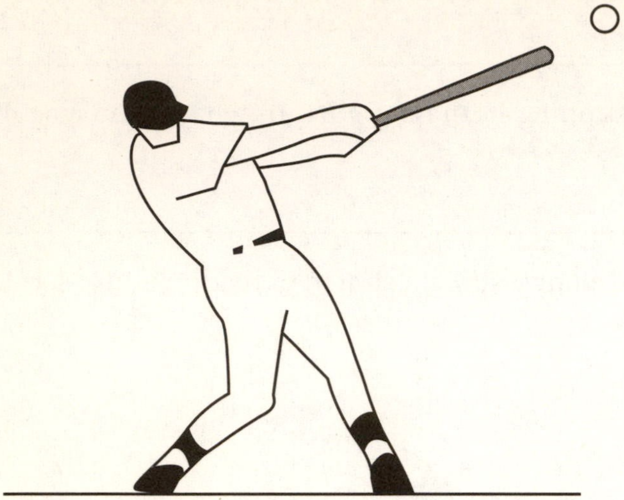

At one moment while in contact with the bat, the bat pushes on the ball with a force of 100 pounds. What is the reaction force?

_____

_____

While in the air, the force of gravity on the ball is 5 ounces. What is the reaction force?

_____

_____

Activity 5.1

Gravity of the Planets

In this activity, you will calculate your weight on all of the planets in our solar system.

The following table contains the mass of the planets as a fraction of the mass of the Earth. It also contains the radius of the planets as a fraction of the Earth's radius.

PLANET	MASS	RADIUS	YOUR WEIGHT
Mercury	0.055	0.38	
Venus	0.86	0.95	
Earth	1.0	1.0	
Mars	0.11	0.53	
Jupiter	320	11	
Saturn	95	9.5	
Uranus	15	4.1	
Neptune	17.2	3.9	
Pluto	0.002	0.24	

The masses are measured in Earth mass units, rather than kilograms. According to the table, Jupiter is 320 times more massive than the Earth. The radii are measured in Earth radius units, so Jupiter has a radius 11 times that of the Earth.

The force of gravity can be calculated by Newton's law of gravitation. Your weight is proportional to a planet's mass divided by the radius of the planet squared. As long as we use the Earth units, it is easy to find your weight on another planet: Multiply your weight on Earth by the mass of the other planet and divide by the other planet's radius squared.

CALCULATE YOUR WEIGHT ON MERCURY

$$W = (\text{weight on earth}) \times \frac{\text{mass of mercury}}{(\text{radius of mercury})^2} = (\qquad) \times \frac{0.055}{(0.38)^2} =$$

Enter this result into the last column in the table.

CALCULATE YOUR WEIGHT ON THE REST OF THE PLANETS

Following the above example, use a piece of scrap paper and a calculator to find your weight on all the planets in the solar system. Enter your results into the table.

ANALYZE YOUR RESULTS

Compare your weight on the Earth to your weight on Venus. It should be a little less than your weight on the Earth. Are you surprised by this result? Explain.

_____

_____

_____

_____

On which planet would you weigh only a little bit more than you do on the Earth?

_____

Does this planet have about the same size as the Earth? What accounts for the fact that you have about the same weight on this planet?

_____

_____

_____

_____

On which planet are you the lightest?

_____

What accounts for the fact that you are lightest on this planet?

_____

_____

_____

_____

Based on the information in the table only, do you think the planets are all made of the same material? Explain.

_____

_____

_____

_____

You're heavier on Neptune than on Saturn, even though Saturn is bigger. Would this still be true if Saturn and Neptune were made of the same material? Explain.

_____

_____

_____

_____

Gravity

TAKING A TRIP TO THE MOON

Suppose you travel in a straight line from the surface of the Earth to the surface of the Moon. At all times you will experience two gravitational forces; the gravitational pull from the Earth and the gravitational pull from the Moon. The Earth's mass is approximately 100 times the Moon's mass.

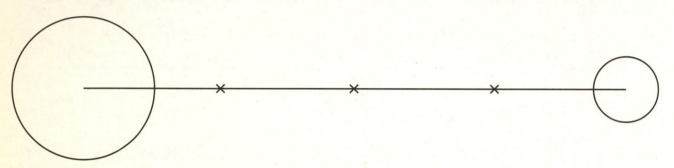

Suppose you are on the line connecting the center of the Earth to the center of the Moon, and you are exactly 1/2 way between them. If the pull of Earth's gravity on you is 1 Newton, what is the pull of the Moon's gravity?

_____

Suppose now that you move along the line until the distance to the center of the Earth is three times the distance to the center of the Moon.

What is the force of gravity on you due to the Earth?

What is the force of gravity on you due to the Moon?

How much closer to the center of the Moon would you have to be, so that the pull of the Moon and the pull of the Earth cancel each other out?

TAKING A WALK ON A RECTANGULAR PLANET

Imagine a hypothetical planet which has rectangular dimensions. This planet is large enough so that you'll feel enough gravity to keep your feet on the ground.

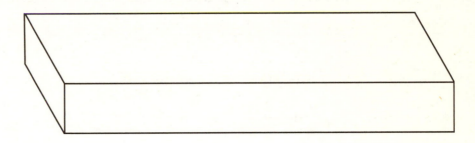

Where would you stand on this planet so that you would feel the lightest? Explain your reasoning and mark the figure with an L at the location.

_____

_____

_____

_____

Where would you stand on this planet so that you would feel the heaviest? Explain your reasoning and mark the figure with an H at the location.

_____

_____

_____

_____

A CYLINDRICAL PLANET

Mark the cylindrical planet with an H and an L for the places where you would feel heaviest and lightest.

Do these locations depend on the relative height of the cylinder?

_____

_____

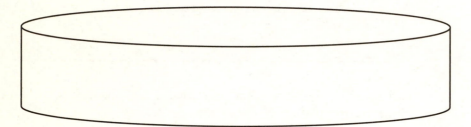

Activity 5.3

Influence of the Sun

Most everyone knows that the Earth executes a near-circular orbit around the Sun. The force of gravity between the Earth and the Sun is responsible for the circular motion.

INVESTIGATE AND RESEARCH

Using your textbook, or any other resources you have available to you, look up the following quantities and write them in the space provided. (Distances should be in meters and masses should be in kilograms.)

The mass of the Sun _____

The mass of the Earth _____

The distance between the Sun and the Earth _____

The mass of your body _____

CALCULATE FORCES

The force between two masses is given by Newton's law of universal gravitation. In equation form, this law reads

$$F = G \frac{m_1 \times m_2}{d^2}$$

Review this law in chapter 5 if you need a refresher. If the masses are in kilograms, and the distance is in meters, the force will be in Newtons. (The gravitational constant G can be found in chapter 5.) Answer the following questions. (Be sure to show your calculations.)

What is the force of gravity between the Earth and the Sun?

What is the force of gravity between you and the Sun?

How many Newtons are there (approximately) in one pound?

If someone pushed on you with a force equal to the gravitational force between you and the Sun, would you feel it? _____

CALCULATE ACCELERATIONS

Newton's second law (chapter 4) allows us to calculate the acceleration due to a net force.
Newton's second law reads $a = \dfrac{F}{m}$.

What is the acceleration of the Earth due to the gravitational pull of the Sun?

What is the acceleration of your own body due to the gravitational pull of the Sun?

What is the acceleration of the Sun due to the gravitational pull of the Earth?

DISCUSSION AND QUESTIONS

The acceleration of the Earth and your own acceleration should be identical. Can you give an explanation of why this *must* be true without discussing forces?

_____

_____

_____

_____

_____

The gravitational pull of the Sun on your own body is certainly not the only force acting on you. But the force we were supposed to use in Newton's second law is the *net force* on you. Why was it OK to treat the gravitational pull of the Sun on your body as the *net force* when calculating your acceleration?

_____

_____

_____

What is the direction of the acceleration of the Earth?

_____

_____

Based on your answer to the last question, why isn't the Earth getting any closer to the Sun?

_____

_____

_____

Activity 5.4

Weightless?

In this activity we are going to calculate the force of gravity on an 80 kilogram person in three different situations. In all three cases you will need to use Newton's law of universal gravitation:

$$F = G \frac{m_1 \times m_2}{d^2}$$

RESEARCH AND COLLECT DATA

Using your textbook, or any other resources you have available to you, look up the following quantities and write them in the space provided. (Distances should be in meters and masses should be in kilograms.)

What is the radius of the Earth? _____

What is the height of Mt. Everest? _____

What is the mass of the Earth? _____

What is the altitude of a typical Space Shuttle orbit? _____

What is the gravitational constant, G? _____

CALCULATE FORCES

Use Newton's law of universal gravitation to calculate the force on an 80 kilogram person in the three different situations. Your answer should be in Newtons.

What is the force of gravity on the 80 kilogram person when he/she is standing on the surface of the Earth?

What is the force of gravity on the 80 kilogram person when he/she is on the top of Mt. Everest?

What is the force of gravity on the 80 kilogram person when he/she is orbiting in the Space Shuttle?

DISCUSSION AND QUESTIONS

In which location is the gravitational force the greatest? Why?

_____

_____

In which location is the gravitational force the least?

_____

Does the fact that the gravitational force is weaker when the person is in the Space Shuttle account for the fact that the person feels *weightless* there? Explain.

_____

_____

_____

_____

_____

Activity 6.1

Momentum: Mass Times Velocity

The momentum of an object is equal to its mass times its velocity. In this activity you will estimate the momentum of a variety of objects. You may have to do some research to find the masses and speeds of the objects described. The mass should be in kilograms, and the speed should be in meters per second.

For reference: A weight of 1 pound is approximately equivalent to a mass of 1/2 kilogram. A speed of 1 mile per hour is approximately equal to 1/2 meter per second.

EXAMPLE: What is the momentum of a sports car driving down the highway?

The speed is probably about 60 miles/hour, or 30 meters per second. The weight of a sports car is about 3000 pounds, or 1500 kilograms.

Momentum=(1500 kg)x(30 m/s)=45,000 kg m/s

1. What is the average momentum of a male sprinter setting a new world record in the 100 meter dash?

2. What is the momentum of a mosquito flying across the room?

3. What is the momentum of the Moon with respect to the Earth? (To determine the speed of the Moon, you will have to find the distance for one complete orbit, divided by the time it takes to orbit the Earth once.)

4. What is the momentum of a major league fastball? (A baseball pitched very fast.)

Activity 6.2

Adding Momentum

Sometimes it is of interest to calculate the momentum of more than one object. The momentum of two or more objects is simply the sum of their individual momenta. Be careful though, because momentum has a direction.

In each of the following exercises, you will be shown two or more masses. Their speed and mass will be indicated, and the direction of the velocity will be indicated by an arrow. For each system, determine the total momentum and draw an arrow to indicate its direction.

1.

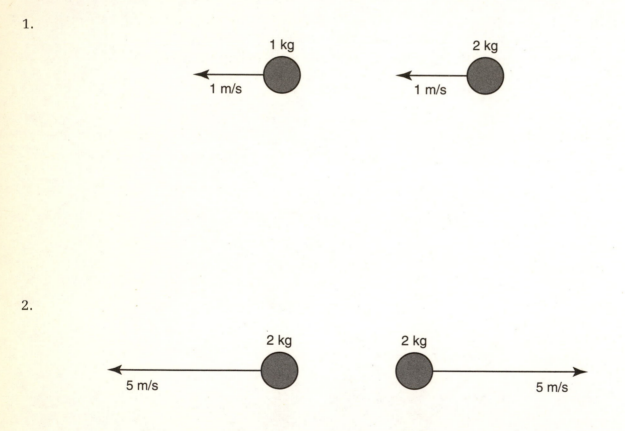

2.

3.

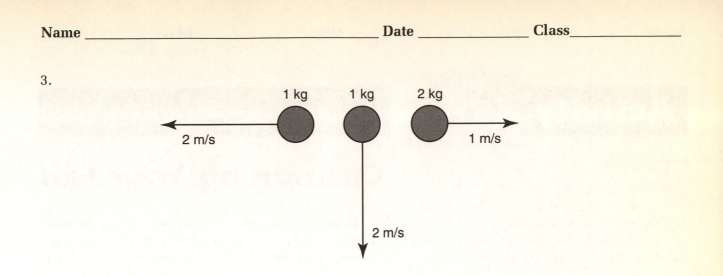

For the following situations, the total momentum may not be exactly vertical or horizontal. In this case, *estimate* the total momentum and draw an arrow in its direction.

4.

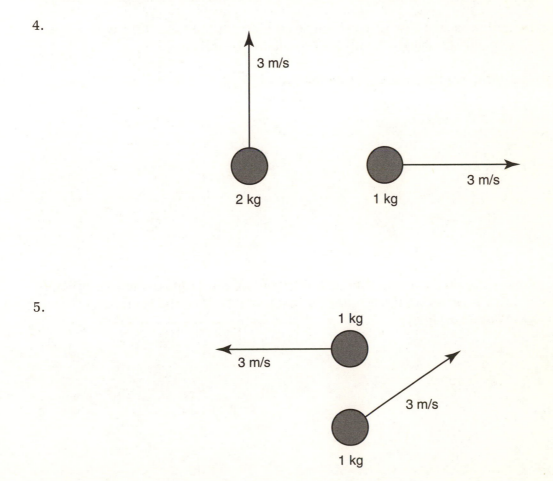

5.

Activity 6.3

Cushioning Your Fall

If you were to accidentally fall off a roof, or high ladder, you would consider yourself very lucky if there happened to be a mattress waiting on the ground to *cushion* your fall. What is it, exactly, about the cushioning in the mattress that makes the fall less dangerous? What is it that hurts you when you fall?

What makes an impact hurt is a combination of the force exerted on you and how that force is distributed over your body. An impact will tend to hurt more if the force is large or if it is distributed over a small area. Football helmets have both a rigid outer shell (which tends to distribute the force over a larger area) as well as cushioning (which tends to reduce the force). In this activity, we will try to determine why cushioning reduces the impact force during collisions.

Consider dropping an egg from a window in your house so that it hits the ground with a speed of 10 meters/second. A typical egg weighs about 7 oz, or about 1/4 kilogram.

What is the momentum of the egg just before it hits the ground?

If the egg hits the concrete sidewalk, it will take about 1/100ᵗʰ of a second to stop completely. What is the force the sidewalk exerts on the egg during this impact? (See the section in chapter 6 titled *Momentum* and *Newton's Laws*.)

Convert your answer to pounds. (1 pound is approximately equal to 4 Newtons.)

Now suppose the egg lands on a fluffy pillow instead of the concrete. Let's assume that the egg takes 1/20th of a second to stop when it lands on the pillow. What is the force the pillow exerts on the egg during this impact?

Convert your answer to pounds.

Do you think the egg will break in either case?

_____

_____

Is the change in momentum greater, less, or the same when the egg lands on the pillow?

_____

What does the cushioning do, exactly, that makes the force less?

_____

_____

_____

_____

_____

Activity 6.4

Conservation of Momentum

According to chapter 6, it takes a force to change an object's momentum. Consider a bottle of water sitting stationary on your kitchen counter. There are certainly forces on this bottle of water. Why doesn't its momentum change? (Review chapter 6 if necessary.)

_____

_____

_____

_____

Let's consider a collision between two hockey pucks on a frictionless ice surface.

When they collide, puck one pushes on puck two and puck two pushes back on puck one. What does Newton's third law tell us about these forces?

_____

_____

_____

If the two hockey pucks are taken together as a single system, do the collision forces change the total momentum of the system? Why?

_____

_____

_____

_____

_____

Chapter 6 discusses *conservation of momentum*. If you slide your textbook across the floor, it will stop due to friction. Clearly the momentum of the book has changed, thus the momentum of the book was not conserved. Under what conditions will the momentum of a system be conserved?

_____

_____

_____

_____

Activity 6.5

Collisions

Consider a collision between two hockey pucks on a frictionless ice surface. Provided there are no external forces on the system (friction or bouncing into the boards etc.), the momentum of the system is the same before and after the collision. In this activity you will be given before and after diagrams of colliding pucks. In some cases they will stick together and in some cases they will not. Fill in the missing information. You will have to use the principle of conservation of momentum. For simplicity, you can assume the pucks have a mass of 1 kilogram.

1.

BEFORE   AFTER

v = ?   v = 0   v = 0   v = 3 m/s

2.

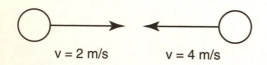

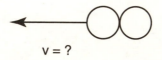

v = 2 m/s   v = 4 m/s   v = ?

3.

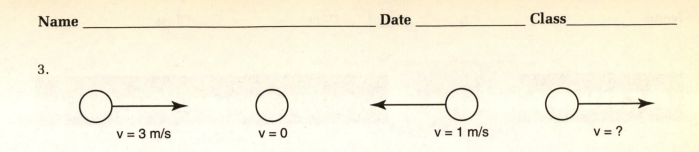

v = 3 m/s   v = 0   v = 1 m/s   v = ?

4.

v = 4 m/s   v = 2 m/s   v = ?

Activity 7.1

Axis of Rotation

Whenever a body rotates, it rotates about a specific *axis of rotation*. The axis of rotation is an imaginary line that passes through points of the body that do not move when the body rotates. A familiar example is the axis of rotation of the Earth. As the Earth spins, points near the equator move in a circular motion. However, imagine an imaginary line connecting the North and South poles. Points along this line do not move as the Earth rotates. This imaginary line defines the axis of rotation of the Earth.

Consider the following diagram of an airplane with three lines drawn through it.

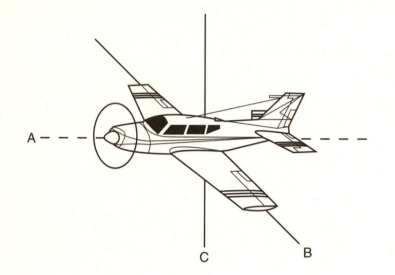

1. If the airplane turned to the right, without banking, which line would best represent the axis of rotation? _____

2. If the airplane banked to the left, without turning right or left, which line would best represent the axis of rotation? _____

3. If the airplane turned upwards, which line would best represent the axis of rotation?

   _____

4. Suppose the airplane simultaneously turned left and banked left. On the diagram, draw the axis of rotation for this maneuver.

Activity 7.2

Period and Frequency

Some objects seem to rotate "faster" than others. In that the Earth rotates once every 24 hours, while the wheels of a fast-moving car may rotate many times in a single second, we may be tempted to say that the wheels of the car rotate "faster" than the Earth. A more precise statement would be to say that the *frequency* of the wheel's rotation is higher than the *frequency* of the Earth's rotation. Frequency is the number of rotations in a given amount of time. The *period* is the time it takes for a complete rotation. Hence, period and frequency are inversely related. A rotation that has a high frequency has a short period.

(The correct adjectives to characterize frequency are *high* and *low*. The correct adjectives to characterize period are *long* and *short*.)

Fill in the blanks in the following sentences:

1. The period of the second hand of a clock is _____ than the period of the minute hand.

2. The frequency of the Earth's rotation about its axis is _____ than the frequency of the Moon's rotation about the Earth.

3. Consider a fast-moving bicycle. The period of the pedal's rotation is _____ than the period of the wheel's rotation.

4. The frequency of occurrence of New Year's Eve is _____ the frequency of occurrence of Thanksgiving.

5. The frequency of occurrence of the Olympic Games is _____ the frequency of occurrence of April Fool's Day.

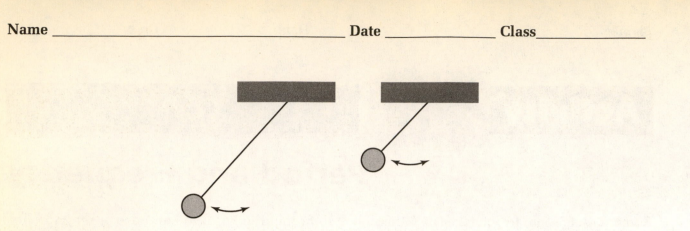

Which do you expect to have a longer period, a short pendulum or a long one? (If you don't know, do an experiment to find out.) Explain.

_____

_____

_____

_____

_____

What has a higher rotation frequency, a rotating compact disc, or the carousel of your microwave oven? (If you don't know, check it out!)

_____

_____

Activity 7.3

Torque

If a tangential force F is applied a distance r from the axis of rotation of an object, then the torque τ is given by $\tau = F \times r$. In other words, torque is the tangential force being applied times the distance from the axis of rotation. The following picture is an irregular object that is free to rotate about an axis which is perpendicular to the page. Five forces, all with the same magnitude, are acting on the object as shown.

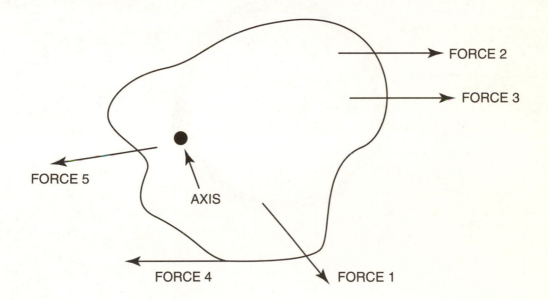

Rank the forces in terms of how much torque they are exerting on the object. The strongest torque would be entered in row 1 and the weakest in row 5.

TORQUE	FORCE
1	
2	
3	
4	
5	

The following picture depicts a bicycle wheel with a variety of forces acting on it. Will the bicycle wheel rotate clockwise or counterclockwise as a result of these forces? Explain your reasoning.

_____

_____

_____

_____

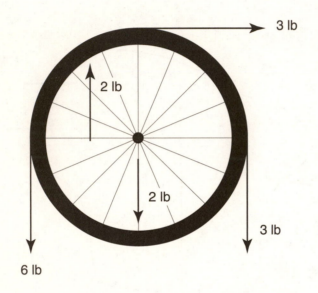

Activity 7.4

Moment of Inertia

Mass is a measure of how much a body will resist changes in linear motion. More mass means it is harder to accelerate that body. *Moment of inertia* is a measure of how much a body will resist changes in rotational motion. Consider a body that is at rest. If it has a large moment of inertia, a large torque will be required to produce a given rate of change in angular speed. If it has a small moment of inertia, less torque is required to produce the same change.

The moment of inertia of a body depends on the mass of the body *and* distribution of mass about the axis of rotation. In general, the further the mass is away from the axis, the higher the moment of inertia.

In the following exercises you will be given a two-dimensional shape. Imagine rotating that shape about the various axes indicated. (Imagine that the axes are perpendicular to the page and you rotate the shapes clockwise or counterclockwise as viewed from above.)

1.

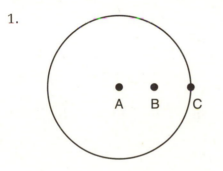

About which axis will the moment of inertia be the greatest? _____

About which axis will the moment of inertia be the least? _____

2.

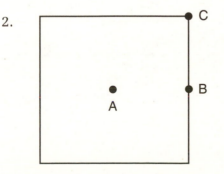

About which axis will the moment of inertia be the greatest? _____

About which axis will the moment of inertia be the least? _____

3.

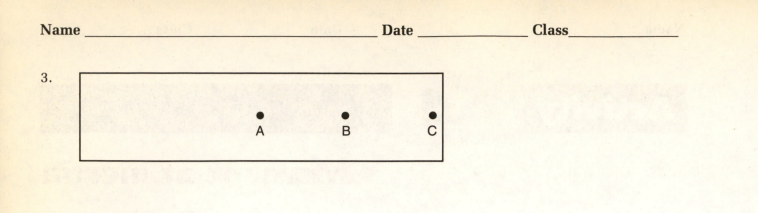

About which axis will the moment of inertia be the greatest? _____

About which axis will the moment of inertia be the least? _____

Two cylinders have the same dimensions (length and diameter) and the same mass. However, one is hollow and the other is solid.

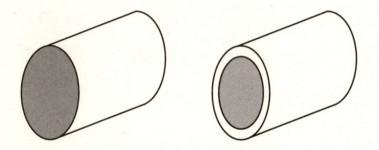

If these two cylinders were released simultaneously at the top of an inclined plane, which one would win a race to the bottom? Explain.

_____

_____

_____

_____

_____

Activity 7.5

Angular Momentum

Consider a bicycle wheel that is free to rotate. A gun that fires wads of putty at high speed is positioned as shown. When the putty hits the tire, it sticks.

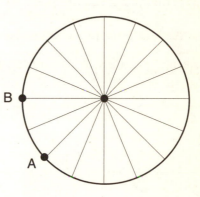

Suppose the tire is initially not rotating.

1. What happens when a wad of putty is fired from and sticks at location A?

_____

_____

2. What happens when a wad of putty is fired and sticks at location B?

_____

_____

3. In which of the above cases does the angular momentum of the wheel change? Does it increase or decrease?

_____

_____

4. A torque is required to change the angular momentum of a system (in this case the wheel). In the case where the angular momentum of the wheel changes, what is the force that creates the torque?

_____

_____

Now suppose that the wheel is initially rotating clockwise.

5. What happens to the moment of inertia of the wheel when a wad of putty sticks to the rim: does it increase, decrease, or stay the same?

_____

_____

6. With the wheel rotating, what happens to the wheel if the putty is fired and it sticks at B?

_____

_____

7. Explain your answer to number 6 in terms of conservation of angular momentum.

_____

_____

_____

_____

_____

Activity 8.1

Work and Power

In this activity you will estimate the power of a crane used to lift steel beams to the top of a tall building. A typical steel beam used in the construction of large buildings has a mass of 1000kg. (1000kg is equivalent to 2200 pounds.)

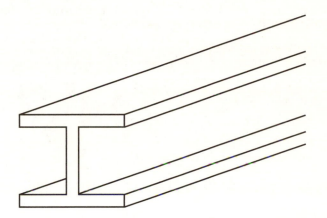

MAKING ESTIMATES

Estimate the height of a tall building, in meters. _____

Estimate the time it should take the crane to lift the beam to the top of the building, in seconds.

_____

WEIGHT OF THE BEAM

The weight of an object is equal to its mass times the acceleration of gravity. For simplicity, let the acceleration of gravity be 10m/s^2. Calculate the weight of the beam. Your answer will be in Newtons.

WORK DONE

Work equals force times distance. If the force is in Newtons and the distance is in meters, the calculated work will be in Joules. Calculate the work done by the crane in lifting the beam.

POWER

Power equals work divided by time. With work in Joules and time in seconds, the calculated power will be in Watts. Calculate the power of the crane.

How many 100 Watt light bulbs need to be lit to equal the power of the crane?

One horsepower is equal to approximately 1000 Watts. What is the power of the crane in horsepower?

Compare the power of the crane to the maximum horsepower of a typical automobile. Are you surprised by your results?

_____

_____

_____

If you wanted to lift the same beams in 1/2 the time, what power motor would you require?

_____

On scrap paper, repeat the calculation of the horsepower of the crane using the following numbers: 100 meter height, 100 second lift time. What is your result? _____

Briefly discuss the reasons for any differences in the two results.

_____

_____

_____

_____

_____

_____

Activity 8.2

Energy and Power

Kinetic energy is given by the formula $KE = 1/2mv^2$. If the mass is in kilograms and the velocity is in meters per second, then the kinetic energy is in *joules*. Thus, a 2 kilogram mass moving at 1 meter per second has a kinetic energy of one joule. ($KE = \frac{1}{2}(2)(1^2) = 1$ joule) It is useful to recall that one kilogram is approximately equivalent to two pounds and that one meter per second is approximately equal to two miles per hour.

In the following questions you will be asked to *estimate* the kinetic energy of various objects moving at various speeds. Be sure to state the mass and the speed that you used in your calculations.

1. Estimate the kinetic energy of a large SUV speeding down the highway.

2. Estimate the kinetic energy of a mosquito buzzing around your room.

3. Estimate the kinetic energy of a sprinter running a world record 100 meter dash.

Recall that 1 Watt is equal to one joule per second. 100 Watts is equal to 100 joules per second.

4. How long would it take a 100 Watt power source to create the energy of the SUV?

5. How long would it take a 100 Watt power source to create the energy of the mosquito?

6. How long would it take a 100 Watt power source to create the energy of the sprinter?

7. Gravitational potential energy is given by the formula $PE = mgh$. About how much gravitational potential energy would you gain by climbing Mt. Everest? (Assume you start your journey from sea level.)

8. How long would it take a 100 Watt power source to lift you to the top of Mt. Everest?

Activity 8.3

Energy and Pennies

We've all heard the story: If you drop a penny off the top of the Empire State Building, the impact will create an eight inch deep hole in the concrete sidewalk. Other people will tell you that it will surely kill anyone that happens to get in its way. Are these just urban myths, or is there any truth to these stories? Surely if these stories are true, we don't want to be walking near tall buildings.

A. Approximately how tall, in meters, is the Empire State Building? _____

B. What is the approximate mass of a penny in grams? Convert this to kilograms.

C. How much gravitational potential energy does a penny gain when it travels to the top of the Empire State Building?

D. Suppose you drop the penny off the top of the Empire State Building and it falls to the road below. Neglecting frictional losses (air drag), what is its kinetic energy just before it hits the road?

E. Neglecting frictional losses, how fast would it be moving just before it hits the road? Convert this speed to miles per hour.

F. Do you think this is fast enough to dig an eight inch hole in the concrete? _____

G. What is the mass of a baseball?

H. How fast would a baseball have to be moving in order to have the kinetic energy calculated in D?

I. Would the baseball in H hurt you if it hit you on the arm? _____

J. Do you think air resistance would play a significant role in a realistic analysis of this problem, or is it acceptable to neglect air resistance? Explain.

Activity 8.4

Conservation of Energy

Consider the simple roller coaster with a single loop-de-loop. A single car starts at A from rest. Neglect any frictional losses.

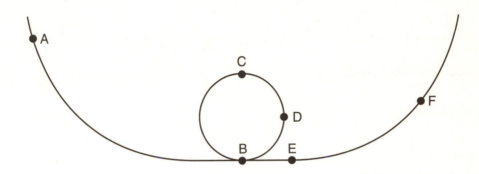

1. Rank the locations indicated on the diagram from highest gravitational potential energy to lowest. If there is a tie between two letters, list those two alphabetically.

   _____

2. There are three locations where there is no work being done on the car by gravity. What are they?

   _____

3. True or False

   _____ The acceleration is the same everywhere.

   _____ The total energy is the same everywhere.

   _____ The speed is the same everywhere.

4. True or False

   _____ The kinetic energy is always increasing.

   _____ The acceleration of the car is zero at E.

   _____ Gravity is doing work everywhere.

Activity 9.1

Avogadro's Principle

Avogadro's Principle states that *equal volumes of a gas at the same temperature and pressure must contain the same number of molecules.* In this activity we will investigate the ramifications of this principle by analyzing some familiar chemicals.

Ammonia is a very toxic and reactive substance. The molecules of ammonia consist of hydrogen and nitrogen atoms. Do some research and find the chemical formula for ammonia.

_____

AMMONIA

Suppose you have a certain amount of ammonia, and you separate the nitrogen and the hydrogen atoms. Nitrogen atoms combine to form gaseous N_2 and hydrogen atoms form gaseous H_2. If the hydrogen gas occupies a volume of six liters, what volume of nitrogen gas is produced?

_____

Which gas has more mass? Explain your reasoning.

Now suppose you have two one gallon containers. One contains water vapor, and the other contains methane gas. The gases are at the same temperature and pressure. Compare the number of molecules in each container.

_____

_____

What is the chemical formula for methane? _____

Which gas has a greater mass? _____

A HYPOTHETICAL EXPERIMENT

You are given an unknown liquid and are asked to determine what atoms the liquid is made of, and the relative masses of the atoms. You separate the liquid molecules and find that you have formed two gases. One gas occupies a volume of three liters and consists of X_2 molecules. The other gas occupies a volume of one liter and consists of Y_2 molecules. The mass of the Y gas is three times the mass of the X gas.

Which of the following could be the chemical formula of the liquid molecules?

a. X_2Y_2

b. XY_2

c. X_6Y_2

d. X_2Y_6

e. X_3Y_2

What is the relative mass of X and Y atoms?

_____

Activity 10.1

Density

Density is a measure of the amount of matter packed into a given volume. Something with lots of mass but a small volume is denser than something that has the same mass but a larger volume. In this activity you will explore the concept of density by approximating the density of an automobile and measuring the density of a book.

For this activity you will need a ruler, an ordinary bathroom scale, and a calculator. Some useful approximate conversions are:

1 kilogram is approximately equivalent to 2.2 pounds.

1 inch is approximately equal to 2.5 centimeters.

THE DENSITY OF AN AUTOMOBILE

What is the density of a typical automobile? To answer this question, you will need to find the approximate mass of an automobile. The mass of an automobile is sometimes printed on the registration certificate or can be found in the owner's manual. If you have no access to a real automobile, then do some research on the internet.

What is the approximate mass of an automobile in kilograms? _____

Now you need to approximate the volume of the automobile. For sake of simplicity, imagine that the automobile has the following shape:

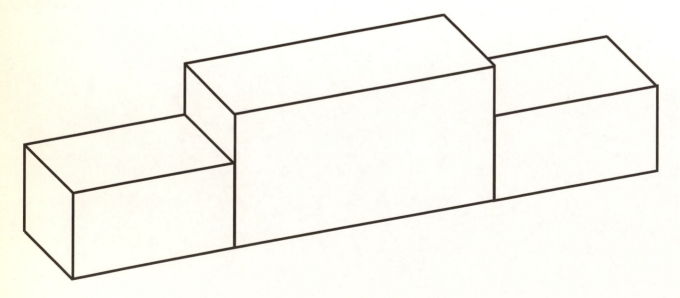

Thus, its volume is simply the sum of the volume of the three boxes. Draw the dimensions of the boxes on the diagram. What is the volume of the automobile in cubic meters? Show your calculations.

Is the density of the automobile greater or less than steel? _____

Does your result make sense? Explain.

_____

_____

_____

DENSITY OF A BOOK

Find a large and heavy book. Weigh it on your scale. What is its mass in kilograms? _____

Measure the length, width, and height of the book. What is its volume in cubic meters? _____

Calculate the density of the book.

Since paper is made from wood, you may expect that the density of a book would be similar to the density of wood. Is the book more or less dense than wood? _____

If there is a difference, can you think of any reasons that may explain the difference?

_____

_____

_____

Activity 10.2

Pressure and Force

Have you ever looked at a glass fish tank and wondered how much the water in the tank weighs? Have you ever wondered how much force is pushing out on the glass walls of the tank? In this activity we will answer these questions. You may need to refer to chapter 10 in the text.

Consider an average fish tank that you may see in someone's home. Its dimensions are 100cm × 50cm × 50cm, as shown.

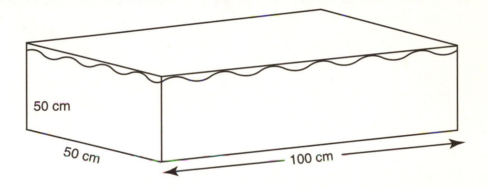

What is the volume of water, in cubic meters, in the tank?

What is the mass, in kilograms, of water in the tank?

What is the approximate weight, in pounds, of water in the tank? (1 kilogram is equivalent to 2.2 pounds.)

Now let's calculate the force pushing out on the front of the tank. Force is related to pressure and area by the relation $F = P \times A$. What is the area of the front panel of the tank, in meters squared? _____

We have the area, but what about the pressure? The pressure is not the same everywhere on the front panel because liquid pressure increases with depth. Luckily, in this case, we can simply use the pressure halfway down the panel and we will get the correct answer. Review the pressure-depth relation, $P = \rho \times d \times g$, in the text.

What is the density of water in kilograms per cubic meter? _____

What is the depth of water halfway down the front panel? _____

What is the acceleration of gravity, g? _____

Calculate the pressure halfway down the front panel of the tank using the pressure-depth relation. The pressure will be in pascals.

Using $F = P \times A$, calculate the force pushing outward on the front panel of the tank. If pressure is in pascals and area is in meters squared, the force will be in Newtons. Convert your answer to pounds.

Activity 10.3

Archimedes' Principle

Any object immersed (fully or partially) in a fluid is subject to a buoyant force. In this activity we will explore the nature of the buoyant force and the rule that allows us to calculate the buoyant force, Archimedes' Principle.

The figure depicts an irregularly shaped object completely immersed in water. The water is pushing in on the object from all sides. At each x in the figure, draw an arrow that represents the force due to the water pressure. The higher the pressure, the longer the arrow.

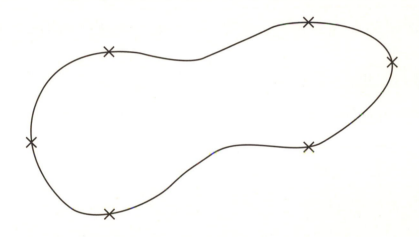

The cumulative effect of the water pushing in from all sides results in a buoyant force that pushes upward. How can you be sure that the buoyant force always pushes upwards?

_____

_____

_____

For an irregularly shaped object, such as the one in the figure, it seems prohibitively complicated to calculate the buoyant force. After all, the water is pushing in from all sides, at all different depths, in all different directions! Thankfully, there once was a man named Archimedes who found that *the upward force exerted on an object immersed in a fluid is equal to the weight of the fluid that the object displaces.*

The figure below shows two tanks that contain fluid. On the left is a tank of water with a person floating in the water. On the right is a tank filled with mercury.

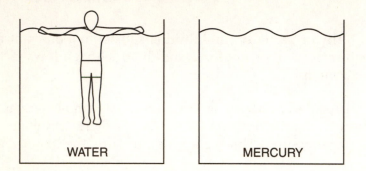

If the person in the water weighs 160 pounds, what is the weight of water that he is displacing?
_____

Convert this weight to kilograms. _____

Since he is almost completely immersed, his volume is about equal to the volume of water displaced. What is the volume of water displaced, in cubic meters?

What is the volume of 160 pounds of mercury?

Suppose he steps out of the water and into the tank of mercury. Draw a stick figure of his body as it would look if he were floating in the mercury. What is the strength of the buoyant force when he is floating in the mercury? _____

Activity 10.4

Gases and Buoyancy

Consider a flexible balloon that is pulled beneath the surface of the water by a strong string.

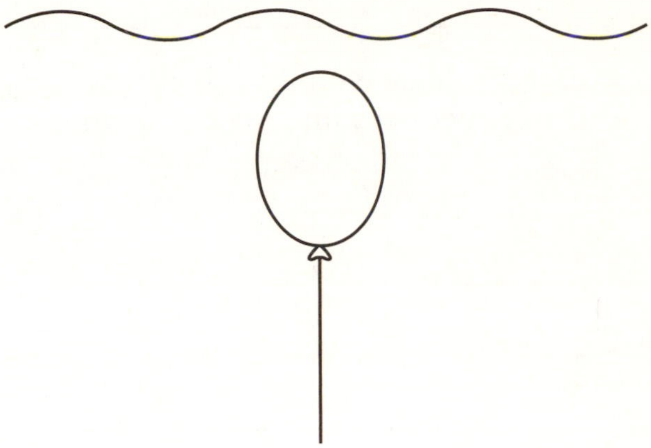

As the balloon goes deeper...

Its volume _____.

Its mass _____.

Its density _____.

The buoyant force on the balloon _____.

Two rocks sit side-by-side at the bottom of a lake. One rock is significantly larger than the other. Which rock, if either, experiences the greater buoyant force? Explain.

_____

_____

_____

A rock is thrown into a pool of water. As it sinks to the bottom, does the buoyant force on the rock change? Explain.

_____

_____

_____

_____

_____

A beach ball and a large rock have the same volume when they are completely immersed in water. Which one, if either, feels a greater buoyant force? Explain.

_____

_____

_____

_____

A rigid steel tank floats in water. The tank is evacuated. Would it float higher or lower in the water if it were pumped full of air? Explain.

_____

_____

_____

_____

A toy boat sails from a pool of water into a pool of alcohol. Does the boat float higher or lower in the alcohol? Does the buoyant force change? Explain.

_____

_____

_____

_____

Activity 10.5

Heating a House of Cubes

In this activity we are going to investigate the laws of scaling and how they are related to real life situations like heating a house. For simplicity, let's assume we have a cubical house. Let its volume be equal to 1 and each face of the cube has an area of 1. Thus the total exposed surface area of the house is 6. (We will assume that the floor is an exposed surface.)

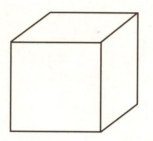

What is the exposed surface area to volume ratio of our cubical house? (Divide the total surface area by the volume.) _____

Suppose we connect two of our cubical houses together to form a duplex, as shown. What is the exposed surface area to volume ratio of the duplex? _____

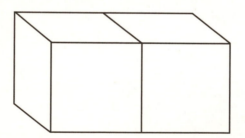

Now suppose that a third house is connected, as shown. What is the exposed surface area to volume ratio now? _____

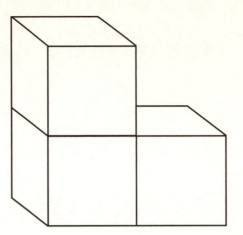

Does it make any difference where the third unit is placed? _____

Based on your results, can you explain why it is more efficient (from an energy efficiency standpoint) to live in an apartment building rather than a single family home?

_____

_____

_____

Construct (draw) the most efficient four unit apartment building you can devise with four cubic units. What is its exposed surface area to volume ratio? _____

Construct (draw) the least efficient four unit apartment building you can devise with four cubic units. What is its exposed surface area to volume ratio? _____

Activity 11.1

The Celsius Scale

Unless you're from the United States, you probably don't need to do this activity. This activity is designed to help familiarize you with the Celsius temperature scale. First, you will be asked to convert some familiar temperatures from Fahrenheit to Celsius. Show your calculations please!

What is body temperature in the Celsius scale?

What is the temperature on a hot day in the Celsius scale?

What is the temperature on a cold day in the Celsius scale?

To what temperature do you cook a turkey in the Celsius scale?

What is the temperature of a frozen lake on the Celsius scale?

Activity 11.2

Applications of Thermal Expansion

The fact that things expand when they are heated has many important ramifications in our everyday life. Indeed, many products we use rely on this fact.

THE BIMETALLIC STRIP

Consider the device shown in the picture. The heart of the device is the bimetallic strip, which is simply two different pieces of metal glued together. An electrical connection can be made if the bimetallic strip bends upward and touches the contact.

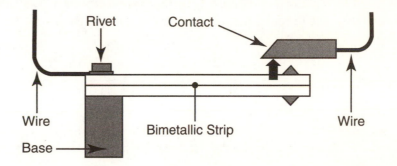

Suppose we want to use this device as a heat detector. When the temperature gets too high, the strip will bend upward and make contact. Which piece of metal making up the strip should have the higher coefficient of thermal expansion? Explain.

_____

_____

_____

NUTS AND BOLTS

The diagram shows an ordinary hexagonal nut. What happens to the area of the hole if the temperature increases?

_____

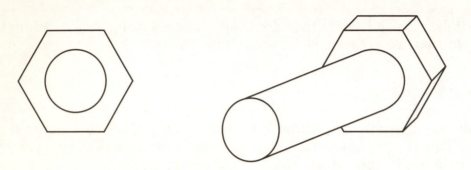

Suppose you screwed a bolt into the nut and the thermal expansion coefficient of the nut was higher than that of the bolt. If the temperature decreased, would the nut be more difficult or easier to remove? Explain.

_____

_____

_____

LIQUID IN A METAL CONTAINER

A steel drum is holding water. The water is filled to the brim. If the temperature increases, water spills over the side. What does that tell you about the relative expansion coefficients of water versus steel? (2 lines)

_____

_____

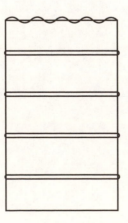

HOW SMALL WILL THINGS GET?

Your friend says that since things get smaller when temperature decreases, they will eventually disappear at low enough temperatures. Is she right? Back up your answer.

_____

_____

_____

_____

_____

Activity 11.3

Calories

Heat is often measured in *calories*, which is the amount of heat necessary to raise the temperature of 1 gram of water by 1 degree Celsius. The familiar food Calorie (with a capital C) is equal to 1000 calories. So one Calorie is equal to the amount of heat (energy) necessary to raise 1000 grams, or 1 kilogram, of water by 1 degree Celsius.

1. Let's start this activity by answering a few preliminary questions.

    a. What is your mass in kilograms? _____

    b. Convert normal body temperature (98.6 degrees Fahrenheit) to degrees Celsius.

    c. A dangerously high fever is associated with a body temperature of about 105 degrees Fahrenheit. What is this temperature in degrees Celsius?

    Let's assume that all of the food that goes into your body is converted into heat energy during digestion. (In reality, only some of the energy liberated during digestion is turned into heat.) Let's also assume your body is made of pure water (not a bad rough approximation).

2. How many Calories of food would you have to ingest in order for your body temperature to become dangerously high?

3. What is the Calorie content of a typical candy bar? How much of that candy bar would be sufficient to raise your body temperature to dangerous levels?

Let's talk about dieting. It seems that everyone knows someone who is trying to lose weight by dieting. One pound of fat stores about 3500 Calories of energy. Thus if your body uses 3500 Calories of energy, you will lose a pound of fat (provided you are maintaining a balanced diet at the time). Your body uses energy when you drink cold water. Why? Because the cold water must be heated to normal temperature by your body. So it's possible (in principle) to lose weight by drinking water.

4. If you drank one liter (1 kilogram) of ice water that was initially at zero Celsius, how much would its temperature increase inside your body? _____

5. How many Calories of energy are necessary to raise the temperature of one liter of water by this amount? _____

6. How many liters of ice water would you need to drink in order to lose a pound?

7. Is this a reasonable way to go about losing weight? Explain.

_____

_____

_____

_____

Activity 11.4

Heat and Internal Energy

The everyday meaning of "heat" is often different from the precise scientific meaning. Heat is the flow of internal energy from one body to another. Review chapter 11 if you need a refresher on this material.

TRUE OR FALSE

_____ An object with a higher temperature always has more internal energy than another object with a lower temperature.

_____ If an object's temperature increases, so does its internal energy.

_____ The only way to increase the temperature of a glass of water is to transfer heat to the water from an object at a higher temperature.

The specific heat of a material is defined as the quantity of heat required to raise the temperature of one gram of that substance by 1°C.

TRUE OR FALSE

_____ Substances with high specific heats tend to cool down faster than substances with low specific heats.

_____ If water had a lower specific heat, it would take less time to heat a cup of water from 10 to 50 degrees Celsius.

_____ You cannot cool an object with a high specific heat to less than 0°C.

Suppose that 500 grams a solid substance, with an initial temperature of 100°C, is dropped into 500 grams of water with an initial temperature of 0°C. Assume that the liquid and solid only exchange heat with each other (the system is thermally insulated).

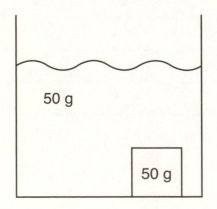

The temperature of the solid will _____. Heat flows from the
_____ to the _____.

Is it possible to determine the final temperature of the mixture? If so, what is the final temperature? If not, what other information would be needed?

_____

_____

_____

_____

A piece of paper and a piece of aluminum foil have the same thickness and are placed in a 400°F oven. After an hour, you open the oven and take out each with your bare hands. The aluminum does not burn your hand, but the paper does.

Is the temperature of the paper likely different than the temperature of the aluminum foil? Explain.

_____

_____

_____

_____

Is the specific heat of paper likely different than the specific heat of aluminum? Explain.

_____

_____

_____

_____

If a solid block of aluminum were in the same oven, you would burn yourself when taking it out of the oven with bare hands. Why?

_____

_____

_____

_____

Activity 11.5

Heat Transfer

RADIANT ENERGY

When it's hot and sunny outside, many people will seek the shade of a tree to cool down. The sun's rays constitute a form of heat transfer called radiation. When you get into a shady area, this radiation is no longer warming your body.

1. Of course it feels cooler under a shady tree, but is the air temperature much cooler under the tree? Explain your reasoning.

   _____

   _____

   _____

   _____

2. Suppose that while sitting outside on a sunny day, your body absorbs 200 Joules of radiant energy from the Sun each second. In order to maintain a constant body temperature, you must be losing 200 joules of heat per second. Explain, in detail, how your body is losing heat though conduction, radiation, and convection.

_____

_____

_____

_____

_____

_____

WIND CHILL

3. On a warm day, a breeze will help keep you cool. This is called the wind chill effect. How does it work?

_____

_____

_____

_____

4. Would the wind chill effect work if the air temperature were greater than your body temperature? Explain.

_____

_____

_____

_____

INSULATION

5. The reason a foam cooler is able to keep beverages and food cool for so long is the low thermal conductivity of its walls. Would a foam cooler be equally effective at keeping hot beverages and food warm? Explain.

_____

_____

_____

_____

Activity 12.1

Interchangeability of Energy

Chapter 12 discusses various forms of energy and the fact that they can be transformed into one another. In the following situations, follow the energy transformations, stating what form of energy is being converted to what other forms of energy. Choose from the following list: *chemical potential energy, gravitational potential energy, elastic potential energy, thermal energy, wave energy, kinetic energy.*

1. A block of wood hangs from a table and falls to the ground, dragging another block across the table.

   _____

   _____

   _____

   _____

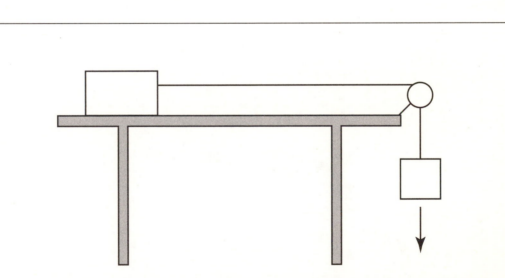

2. A rock is dropped into a small pond and creates a wave that breaks on the shores of the pond.

   _____

   _____

   _____

   _____

3. An archer draws his bow and fires and arrow towards the top of a tall tree. The arrow sticks into the tree.

_____

_____

_____

_____

4. A stick of dynamite explodes, creating a fireball and a large sound.

_____

_____

_____

_____

5. A woman steps on the gas to power her car up a steep hill.

_____

_____

_____

_____

6. The sun sends out radiation that heats the water in your solar powered house.

_____

_____

_____

_____

Activity 12.2

The First Law of Thermodynamics

The first law of thermodynamics states that *in a closed system, the change in internal energy equals the heat added to the system minus the work done by the system.*

In the following situations there is an increase in the internal energy of the system. State whether the change in internal energy is a result of work done by the system or heat added to the system.

The internal energy of water increases when it is put on a hot stove. _____

The internal energy of your hands increases when you rub them together. _____

The internal energy of the air in your bicycle tires increases when you sit on your bicycle.

_____

The internal energy of the water in your swimming pool increases as spring becomes summer.

_____

THINKING MORE ABOUT WORK

Suppose an ordinary party balloon is moved from a cold room to a warm room. Heat is conducted through the rubber balloon and the temperature of the gas inside increases. The balloon expands and the temperature of the gas in the balloon increases.

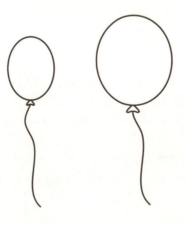

1. The balloon did work in the process of expanding. We learned in previous chapters that work is a force times a distance. What is the force? What is the distance?

   _____

   _____

   _____

2. If the heat added was 500 joules, and the change in internal energy was 900 joules, how much work did the balloon do?

3. Suppose the balloon had been rigid, so it didn't expand during the heating process. Would the change in internal energy be greater or less than in the case with the expanding balloon?

   _____

   _____

4. If the balloon were rigid, would the temperature increase be greater or less than in the case with the expanding balloon?

   _____

   _____

5. How much work did the balloon do in this case?

Now consider a piston and cylinder that contains air. Let's think about what happens when the piston is pushed into the cylinder so the gas is compressed. Suppose the force pushing on the piston is approximately 300 Newtons and the piston moves 1 centimeter (0.01 m).

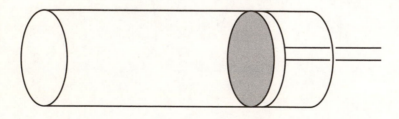

6. Calculate the work done.

7. If 2 joules of heat was lost from the cylinder during the compression, what was the change in internal energy?

Activity 13.1

The Second Law of Thermodynamics and Perpetual Motion

Chapter 13 presents three different statements of the second law of thermodynamics. They are (1) *heat does not spontaneously flow from a cold body to a hot body;* (2) *you cannot construct an engine that does nothing but convert heat completely into useful work;* and (3) *every isolated system becomes more disordered with time.*

1. A refrigerator causes heat to flow from a cold beverage to the warm air outside. Why doesn't this contradict (1)?

   _____

   _____

   _____

   _____

2. The heat generated in the cylinder of your car engine is converted into work as your car is accelerated up a hill. Why doesn't this contradict (2)?

   _____

   _____

   _____

   _____

3. After cleaning up your room, it seems more ordered. Why doesn't this contradict (3)?

   _____

   _____

   _____

   _____

In watching a movie of each of the following situations, could you determine whether or not the movie was being run in reverse? Yes or no?

A ball flies through the air   _____

An ice cube melting   _____

An apple falling towards the ground   _____

An oil slick spreading on the water   _____

A croquet ball rolls to a stop on a grass court   _____

A perpetual motion machine is sometimes referred to as any device that violates the second law of thermodynamics. Consider the following perpetual motion machines and explain why they don't work. You may want to do a little research if you can't figure out the trick.

THE BHASHKARA MACHINE

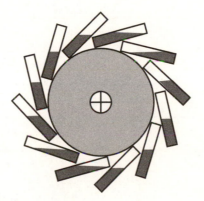

This is a wheel with containers of mercury around its rim. As the wheel turns, the mercury is supposed to move within the containers in such a way that the wheel would always be heavier on one side of the axle. How come this machine does not work?

_____

_____

_____

_____

_____

SELF FLOWING FLASK

How come this "self flowing flask" does not continually circulate water?

_____

_____

_____

_____

_____

THE OVERBALANCING WHEEL

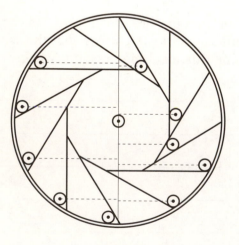

The rolling balls are supposed to pull down on the left more than they pull down on the right, causing a perpetual counterclockwise rotation of the wheel. How come this ancient design does not work?

_____

_____

_____

_____

_____

Activity 13.2

Entropy and Probability

Entropy and probability are intimately related. In this activity we will look at the relationship between the number of different arrangements of a system (the disorder) and the probability that the system will be in one particular state. You will need a calculator for this activity. Review chapter 13 as necessary.

1. How many different ways can five white balls and one black ball be arranged? (balls of the same color are indistinguishable) _____

2. How many different ways can four white balls and two black balls be arranged?

_____

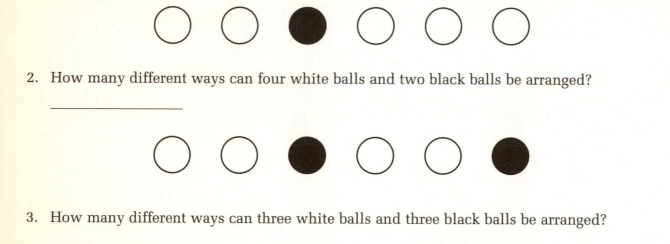

3. How many different ways can three white balls and three black balls be arranged?

_____

4. What is the probability that a random arrangement of the balls in number 3 will yield the combination shown below? _____

Now consider a 4x4 checkerboard with eight black checkers.

5. How many different ways can the eight checkers be arranged on this checkerboard?

6. What is the probability that placing the checkers on the board randomly will yield an "ordered" state where all of the checkers are on the left half of the board?

7. How many different ways can the checkers be arranged so that four are on the left half of the board and four are on the right half?

8. What is the probability that placing the checkers on the board randomly will yield a state where four checkers are on the left half of the board and four are on the right half?

9. How do your answers to 5-8 relate to entropy and the second law of thermodynamics?

_____

_____

_____

_____

_____

Activity 13.3

Monte Carlo

In chapter 13 you learned that systems tend to become disordered as time goes by. Thus, the likelihood of finding an ordered state in nature can be quite small. In this activity we will demonstrate this with a Monte Carlo Simulation. Generally speaking, a Monte Carlo Simulation is the use of a random number generator to simulate the behavior of a model.

The model we would like to simulate is a *random walk* in one dimension. Consider the following line of five X'es. Place a marker on the middle X (a button or paper clip will work just fine). Now flip a coin. If you get heads, move the marker to the right, tails move it to the left. Keep flipping the coin until your marker lands on the left most X. If the coin is on the right most X and you flip heads, you don't move the coin, but it still counts as flip.

1. How many flips of the coin did it take before the marker reached the left most X?

   _____

2. Repeat the simulation. How many flips did it take this time? _____

Now look at the array of numbers on page 109. These numbers were generated randomly and range from 0 to 9. They will serve as a random number generator for our next simulation, just as flipping a coin serves as a random right-left generator. In order to generate a sequence of random numbers from 0-9, just start anywhere on the grid and work your way up-down or left-right.

Consider the 10 spaces numbered 0 to 9. Place five markers (coins will work well) on five different locations as follows:

• Pick a random number.

• Place a marker on that number if it is not already occupied.

• Repeat until all five markers are placed.

Now we will simulate the random walk as follows:

- Pick a random number.

- If that marker is occupied, flip a coin. Heads means a move to the right, tails means a move to the left.

- Move the marker right/left only if the space is unoccupied. Also a marker on space 9 cannot hop to the right and a marker on space 9 cannot hop to the left.

- Repeat this procedure until (1) all five markers are in spaces 0-4, or (2) you run out of random numbers.

3. What was the outcome of your simulation?

_____

_____

_____

4. How is this simulation related to entropy and the second law of thermodynamics as discussed in chapter 13?

_____

_____

_____

_____

_____

```
2 2 1 6 8 6 9 0 0 9 5 0 0 4 0 6 3 4 0 9 2 8 1 1 4 5 4 9 9 3
6 1 2 4 1 1 3 7 0 7 3 4 7 8 6 1 2 4 8 1 4 0 3 6 4 6 1 9 7 0
3 9 8 8 5 7 3 1 0 5 0 5 7 1 4 9 3 7 4 4 2 8 5 4 5 5 7 0 1 6
4 6 0 3 4 7 5 2 3 3 1 6 4 7 4 5 4 4 7 2 8 8 5 4 9 0 6 8 9 8
0 6 1 7 8 0 3 1 2 5 1 2 0 2 8 5 2 3 8 2 6 4 8 9 2 3 5 8 9 7
2 8 9 7 9 7 9 2 9 9 6 0 6 9 0 2 0 1 0 0 2 6 6 0 8 3 2 3 8 7
6 8 6 0 5 6 5 4 6 5 1 3 3 9 6 2 8 3 3 1 8 6 2 7 4 4 8 0 6 8
9 3 5 4 8 5 4 2 8 0 0 6 0 1 8 7 5 5 1 0 6 0 8 7 7 9 5 3 6 4
8 7 2 8 1 0 6 0 9 3 9 7 8 7 7 0 3 2 0 8 0 9 3 3 3 9 3 0 4 7
5 4 5 4 9 8 6 5 5 5 9 5 9 0 9 3 3 1 2 2 8 6 8 1 6 1 2 5 0 7 9
6 6 2 4 8 5 1 6 3 8 0 5 7 3 6 1 5 4 0 4 6 0 4 0 7 1 1 0 0 5
6 8 6 0 6 0 4 4 4 4 0 7 7 7 7 3 1 5 6 6 5 5 6 2 8 8 0 8 6
4 2 1 3 0 3 6 1 6 6 9 1 7 7 5 9 2 0 8 0 3 5 9 0 8 9 9 7 8 7
1 7 2 7 9 3 5 8 1 7 1 8 1 1 0 9 0 7 5 7 5 3 4 7 4 7 6 1 7 6
6 3 4 7 9 4 5 8 4 1 9 9 1 2 0 0 1 5 7 3 9 1 4 0 9 4 4 6 1 3
3 2 5 4 6 9 8 1 0 7 6 6 8 5 1 7 6 1 5 6 1 3 4 8 0 5 8 3 7 6
2 3 4 6 5 3 9 0 1 3 4 7 2 8 4 6 2 0 3 9 4 0 6 6 9 8 2 7 4 3
0 6 5 3 3 2 5 3 7 4 2 0 9 8 8 2 0 8 2 3 0 6 1 1 6 7 6 8 1 1
7 6 4 0 2 8 6 3 4 0 1 0 9 1 6 6 0 8 0 1 5 1 8 6 5 4 7 2 3 3
7 2 7 0 3 2 2 9 4 3 2 7 6 3 6 0 9 9 5 5 8 2 6 7 4 5 9 0 9 7
1 5 6 2 8 8 6 2 9 4 3 9 9 0 3 9 8 4 2 5 4 2 7 5 3 0 5 3 4 3
5 3 1 0 7 0 2 0 0 3 6 6 2 1 6 5 6 6 1 0 5 6 0 6 4 2 7 4 2 7
0 6 1 4 2 2 4 8 1 0 9 4 1 2 2 7 9 6 9 4 1 9 1 5 9 2 3 9 8 7
6 8 8 1 6 7 8 3 0 1 4 5 5 1 8 1 5 6 1 8 5 7 5 1 2 8 2 1 2 4
7 6 0 3 9 8 6 3 4 7 2 2 6 5 2 9 7 5 9 0 1 3 7 7 7 6 8 3 0 6
5 1 3 9 9 5 7 1 7 4 9 0 2 2 0 8 6 0 9 2 0 0 4 1 3 5 1 5 4 5
0 1 7 0 0 3 3 2 2 6 1 5 6 2 7 3 1 3 5 1 0 6 0 6 0 0 7 4 3 3
0 3 6 7 9 4 0 9 9 6 1 8 5 4 7 2 0 6 3 1 5 2 2 9 6 8 6 0 2 3
8 2 7 7 4 1 7 9 8 4 3 3 6 5 7 2 8 1 8 3 3 2 2 1 9 8 1 8 8 7
3 9 4 9 1 5 0 0 3 6 3 1 1 5 8 4 2 5 6 0 8 5 5 6 2 9 9 8 5 3
9 6 2 7 7 7 6 5 4 6 0 5 7 8 3 9 2 2 2 3 0 6 6 4 8 8 2 0 4 5
3 9 0 8 9 9 3 4 5 9 3 4 3 6 2 3 6 0 4 9 8 5 1 4 8 1 5 6 7 7
5 3 2 9 3 0 4 0 8 1 3 7 4 4 2 1 9 1 3 4 3 9 8 9 0 0 7 0 2 2
7 9 2 1 3 7 7 2 1 5 5 1 8 2 2 2 5 2 8 2 6 4 2 6 6 6 1 2 0 6
2 4 5 3 7 8 8 2 0 6 6 1 3 2 5 4 8 9 3 9 9 1 1 1 4 5 2 1 6 0
5 1 3 4 7 4 3 9 1 5 5 2 3 4 2 1 3 7 8 6 5 9 9 5 6 5 8 9 3 0
7 2 6 6 6 3 9 5 4 4 9 8 9 4 2 2 4 6 3 4 0 1 9 7 6 1 0 1 5 0
4 4 1 9 3 9 3 1 8 1 7 2 1 0 3 6 4 5 2 0 0 7 6 8 0 6 9 1 1 2
7 3 6 0 2 6 8 2 0 3 2 6 5 4 8 8 5 3 0 7 2 7 9 6 7 5 6 7 8 2
9 8 0 6 0 9 4 4 5 9 3 6 2 2 0 6 8 0 6 0 6 7 4 4 3 4 7 9 1 0
6 4 0 0 0 3 8 5 5 2 4 6 7 5 1 7 0 5 4 9 6 2 9 1 2 4 3 2 8 8
5 6 0 8 6 4 9 9 9 6 0 2 8 3 5 1 8 9 3 7 2 7 7 3 0 6 7 7 0 2 4
2 3 7 6 2 0 5 6 1 0 0 6 9 8 0 0 9 1 2 3 1 4 1 8 4 8 2 5 3 2
1 2 9 4 9 5 9 3 2 8 8 2 1 3 2 5 2 0 2 0 7 7 5 7 3 5 7 8 7 3
1 4 0 2 7 3 2 8 4 2 7 4 6 2 0 9 3 8 3 7 7 9 5 2 1 5 7 4 1 4
6 9 7 5 7 3 9 7 9 9 7 9 1 4 7 4 4 4 2 4 2 3 1 6 0 9 7 0 8 7
8 8 8 2 6 0 1 2 3 6 5 0 8 1 6 1 4 6 4 8 0 1 9 4 8 7 0 5 0 5
6 4 3 1 8 8 6 3 3 9 6 5 1 6 3 5 0 8 2 4 8 5 9 9 7 9 3 4 6 6
9 8 2 6 0 2 3 0 2 0 1 5 8 7 6 2 8 7 5 1 8 5 1 7 9 8 9 1 5 3
5 4 6 3 2 8 4 7 6 7 7 5 0 3 7 8 1 8 7 0 0 4 8 3 3 9 8 5 8 7
3 1 4 1 7 0 5 1 6 7 4 4 7 0 2 7 0 4 5 9 9 1 8 9 8 7 6 2 8 5
5 7 8 1 4 6 9 6 2 9 5 2 6 6 6 5 0 1 2 7 6 0 3 9 2 7 1 1 0 1
0 8 9 4 9 3 4 2 5 8 3 9 6 4 2 9 2 5 1 4 7 5 3 4 3 8 2 7 8 5
4 7 7 4 2 9 1 1 1 6 8 1 2 5 3 3 4 4 6 7 9 2 1 2 1 6 7 4 0 8
4 1 0 7 4 7 7 0 6 1 4 1 9 6 8 2 1 2 2 5 9 4 3 0 7 7 6 3 7 2
8 3 5 8 3 5 7 2 6 8 1 5 6 3 4 4 0 3 9 4 8 3 7 9 0 9 2 1 6 0
2 5 5 7 2 3 9 2 0 5 1 0 4 1 2 4 4 5 9 1 0 4 2 8 2 8 5 6 8 5
2 0 4 6 9 5 3 0 3 2 0 9 3 8 3 7 3 9 3 7 2 5 9 0 7 1 4 9 9 6
3 8 3 5 8 8 0 3 6 8 9 5 4 7 0 5 7 2 8 8 1 2 8 1 1 2 0 6 7 6
2 6 1 2 1 3 9 5 1 7 4 6 4 7 4 1 9 2 7 2 6 6 9 6 8 3 9 1 1 9
4 5 9 9 1 4 2 8 9 2 2 3 4 6 7 4 3 6 0 4 2 8 6 8 6 9 6 7 9 5
1 2 4 2 0 9 1 0 5 4 3 1 9 5 5 7 5 5 5 4 8 6 3 0 0 4 5 3 0 0
8 5 5 6 3 8 5 0 0 3 4 3 7 0 0 0 1 3 3 1 6 8 9 6 5 5 0 5 8 8 1
2 4 6 3 2 2 8 4 5 8 8 1 5 2 1 0 3 0 8 5 9 5 9 6 3 7 6 7 3 8
8 7 5 4 2 5 9 3 2 7 5 2 5 9 4 7 1 8 7 5 0 3 8 0 8 5 0 0 9 5
9 9 3 0 6 0 5 4 9 4 2 7 4 2 5 0 2 3 3 3 7 9 4 6 5 6 8 9 2 0
1 0 0 2 5 4 7 3 7 6 5 3 9 9 1 3 3 3 3 7 5 6 7 1 9 4 9 6 2 6
6 3 8 4 6 6 7 9 4 9 9 1 1 2 0 5 9 0 0 9 9 0 2 4 4 3 5 1 4 8
7 6 4 7 4 9 9 1 1 6 6 6 4 6 3 2 4 3 8 2 1 1 4 3 4 7 8 5 2 0
```

Activity 14.1

A Pendulum Experiment

A simple pendulum is a system that executes simple harmonic motion. In this activity you will investigate the relationship between the period, frequency, and length of a simple pendulum. You will need some string, a ruler, a small mass that you can tie the string to, and a stopwatch or watch with a second hand. The mass should be heavy enough so that the pendulum swings back and forth freely for many cycles before stopping.

Describe the string you are using.

_____

_____

Describe the mass you are using.

_____

_____

For five different lengths of string, measure the period and frequency of the pendulum. The best way to measure the period is to determine the time it takes the pendulum to swing back *and* forth 10 times. Divide this time by 10 to get the period. Fill in the table with your data. Be sure to include the units of your measurements.

LENGTH	PERIOD	FREQUENCY

Plot the period versus the length on the graph below.

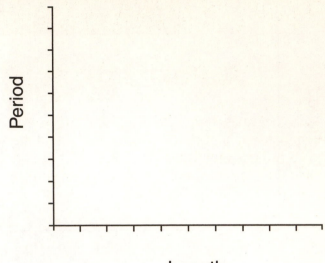

Plot the frequency versus length on the graph below.

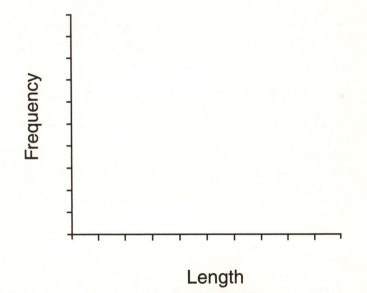

Describe how the period and frequency of a pendulum are related to the length.

_____

_____

_____

_____

In the graph below, plot the period versus the frequency.

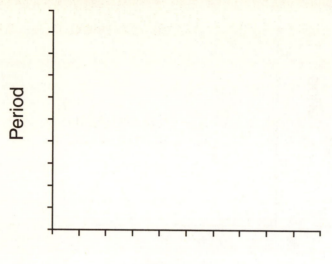

What kind of relationship is this?

_____

Is there an equation in the book that confirms your conclusion? If so what is it?

_____

_____

Activity 14.2

Waves

Consider the following "snapshot" of a wave. It could be a water wave or a wave on a string. Use a ruler to measure the dimensions of this wave in centimeters.

1. What is its wavelength? _____

2. What is its amplitude? _____

3. If the wave is traveling at a speed of 3 cm/s, what is its period?

4. What it its frequency?

When two waves encounter each other, they *interfere*. The height of the resultant wave is simply the sum of the heights of the interfering waves. Consider the two wave pulses pictured below. They are traveling toward each other at a speed of one centimeter per second and their peaks are separated by a distance of 4 centimeters. Sketch the resultant wave 1, 2, 3, and 4 seconds later.

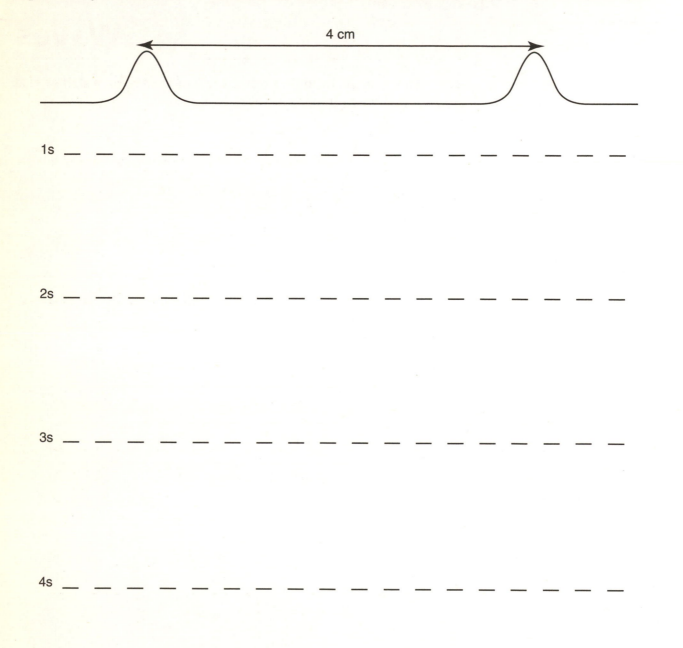

1s —

2s —

3s —

4s —

Consider the two wave pulses pictured below. They are traveling toward each other at a speed of one centimeter per second and their peaks are separated by a distance of 4 centimeters, except this time the right wave is inverted. Sketch the resultant wave 1, 2, 3, and 4 seconds later.

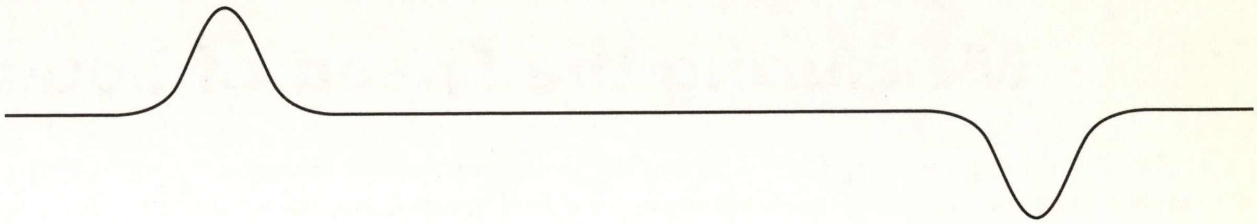

1s _ _ _ _ _ _ _ _ _ _ _ _ _ _ _ _ _ _

2s _ _ _ _ _ _ _ _ _ _ _ _ _ _ _ _ _ _

3s _ _ _ _ _ _ _ _ _ _ _ _ _ _ _ _ _ _

4s _ _ _ _ _ _ _ _ _ _ _ _ _ _ _ _ _ _

Activity  15.1

Measuring the Speed of Sound

The speed of sound can be measured by doing the following simple experiment. You will need a stopwatch and as long a measuring tape as you can find and two blocks of wood to clap together. If blocks of wood are not available, anything that makes a sharp clapping sound will do. You will need a friend or classmate to help with the experiements.

Find a building that has a large, unobstructed outside wall. Stand a measured distance away, facing the wall. If the blocks are clapped together, an echo should be heard as the sound travels to the wall and is reflected back. The trick to the experiment is to synchronize the claps with the echoes so that you clap at precisely the time the echo from the previous clap reaches you. The time between claps is equal to the time it takes for the sound to travel to the building and back.

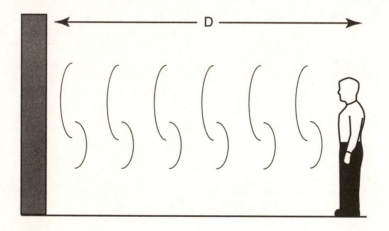

Find a location in front of the building and measure the distance to the building. What is the distance to the building? _____

Start clapping and get a feel for synchronizing the claps with the echoes. Once you get a feel for it, have a friend time 10 claps. The time between claps is this time divided by 10.

What is the time for 10 claps? _____

What is the time between consecutive claps? _____

Calculate the speed of sound. Show your work.

How close is your answer to the speed of sound presented in chapter 15?

_____

_____

Think of several possible sources of error in this experiment and discuss how they may have affected your results.

_____

_____

_____

_____

_____

Another way to measure the speed of sound would be the following: Two people stand far from each other on a flat, open space. One claps the blocks, the other starts a stopwatch when the blocks are clapped and stops it when the sound is heard. The speed of sound is the distance between the people divided by the time. Do this experiment.

What is the distance between the people? _____

What is the time? _____

Calculate the speed of sound. _____

Think of several possible sources of error in this experiment and discuss how they may have affected your results.

_____

_____

_____

_____

_____

Compare the two methods. Which do you think gives a more accurate result? Why?

_____

_____

_____

_____

_____

_____

Activity 16.1

Positive and Negative Charge

For the most part, the world is made up entirely of three elementary particles: electrons, protons, and neutrons. The nucleus of the atom contains protons and neutrons and electrons orbit the nucleus.

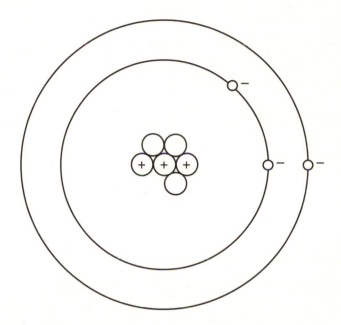

Neutrons have no electrical charge, protons have a positive charge, and electrons have a negative charge. Usually there is the same number of electrons and protons in matter, which deems it electrically neutral. Like charges repel and opposite charges attract.

In this activity we will experiment with static electricity. You will need several ordinary party balloons and some thread.

HAIRBALLOON

Blow up a balloon and rub it in your hair vigorously. Then hold it near your head. Describe what happens.

_____

_____

_____

_____

Explain what happened in terms of the movement of electric charge from the balloon to your hair or vice versa.

_____

_____

_____

_____

HANGING BALLOONS

Now blow up two balloons, rub them in your hair, then hang them from threads.

Describe what happens.

_____

_____

_____

_____

Do they repel or attract? Explain what happened in terms of electrical forces between the balloons.

_____

_____

_____

_____

ROLLING A CAN WITH A CHARGED BALLOON

Find an empty aluminum soda can. Place it on a flat, smooth surface. Charge a balloon by rubbing it in your hair. It should be possible to roll the can along the surface by bringing the balloon close enough.

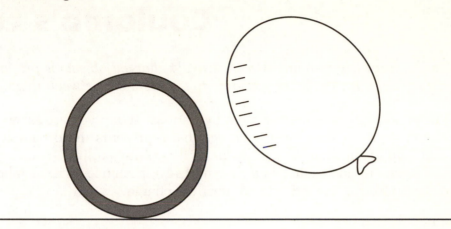

Were you able to roll the can? _____

The balloon was negatively charged. In the figure, the negative charge is indicated by the – signs. Draw + and – signs on the can to show how the charge is distributed on the can when the balloon is close by.

Would this experiment have worked better if the balloon had been charged positively? Explain.

_____

_____

_____

_____

DEFLECTING A WATER STREAM

In a bathtub or sink create a very thin, smoothly flowing stream of water from the faucet. Hold a charged balloon near the stream and see what happens. Was the stream deflected?

_____

The water is not charged. How is it that neutral water can be deflected by a charged balloon?

_____

_____

_____

_____

Activity 16.2

Coulomb's Law

Coulomb's law states that *the force between any two electrically charged objects is proportional to the product of their charges and inversely proportional to the distance between them.*

In this activity you will be given a number of charge configurations. When you are asked for the force on a particular charge, draw an arrow at the charge that represents the direction of the force. The magnitude and sign of the charges will be given in arbitrary units. (A charge of -2 has twice the magnitude and the opposite sign as a charge of +1.) Distances will also be given in arbitrary units. If the force is zero, just write "zero" near the charge.

1. What is the force on the +2 charge?

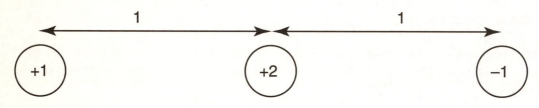

2. What is the force on the left most charge?

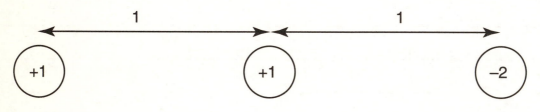

3. What is the force on the +2 charge?

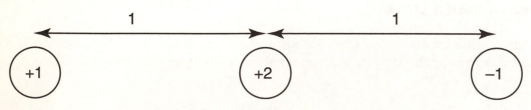

4. What is the force on the middle charge?

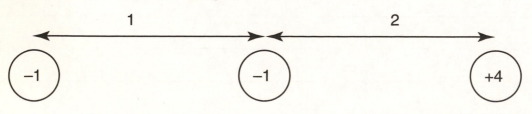

The following charges reside on the corners of a square.

5. What is the force on the -2 charge?

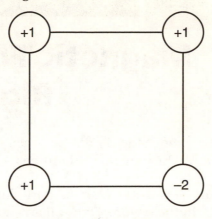

6. What is the force on the lower +1 charge?

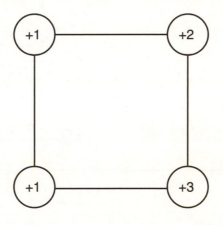

7. What is the force on the +2 charge at the center of the square?

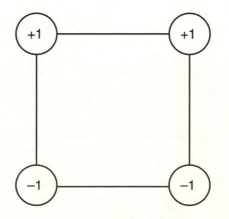

Activity 16.3

Magnetic Forces and the Right-Hand Rule

When a charged particle moves in a magnetic field, it experiences a force. The direction of the force is given by the *right-hand rule*. Review this rule in chapter 16.

In the following exercises you will be shown a charged particle. Its velocity, v, will be indicated as well as the magnetic field, B, in the region. Draw an arrow on the charge that represents the direction of the force. If the force is into or out of the page, just write "into page" or "out of page."

1. A negatively charged particle moves to the right. The magnetic field points into the page.

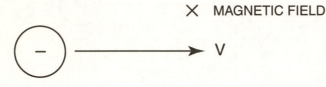

2. A positively charged particle moves up. The magnetic field points to the left.

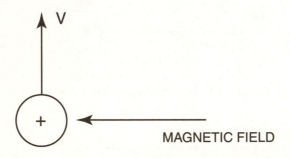

3. A negatively charged particle moves into the page. The magnetic field points to the left.

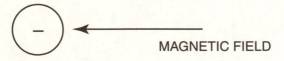

4. A positively charged particle moves out of the page. The magnetic field points down.

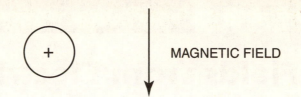

5. A negatively charged particle moves down. The magnetic field points to the left.

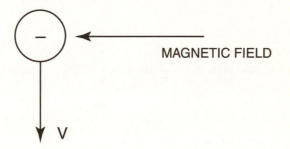

6. A positively charged particle moves to the left. The magnetic field points down.

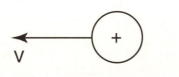

Activity 17.1

Magnetic Fields from Electric Current

In chapter 17 you learned that moving electric charges create magnetic fields. In the following exercises you will be asked to determine the direction of the magnetic field created by various electric currents.

1. The figure represents a long straight wire carrying current into the page. Draw arrows at A, B, C, D, and E that represent the direction and relative strength of the magnetic field at those locations. (Longer arrows mean stronger fields.)

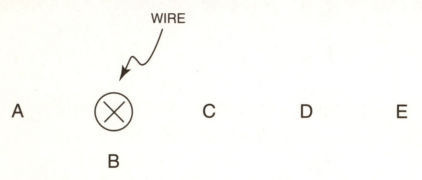

2. A circular loop of wire carries current counterclockwise as viewed from above. Draw arrows at A, B, C, D, and E that represent the direction and relative strength of the magnetic field at those locations. (These locations are in a plane perpendicular to the plane of the loop and bisecting the loop.)

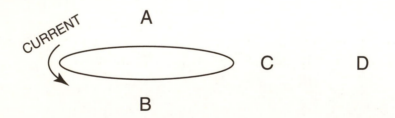

3. Two long straight wires carry current of equal magnitude in opposite directions, as shown.

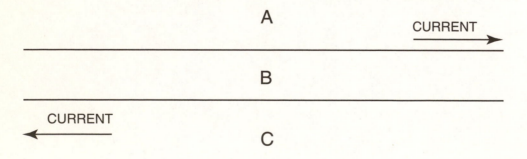

Select the location, A, B, or C, where

the magnetic field is the strongest _____

the magnetic field is zero _____

the magnetic field points up (out of the page) _____

the magnetic field points down (into the page) _____

4. Two circular loops carry the same amount electric current.

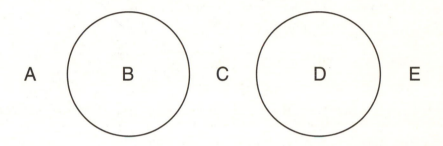

Suppose the current runs clockwise as viewed from above in both loops. What is the direction of the magnetic field at A, B, C, D, and E?

A _____

B _____

C _____

D _____

E _____

Suppose the current runs clockwise in the left loop and counterclockwise in the right loop. What is the direction of the magnetic field at A, B, C, D, and E?

A _____

B _____

C _____

D _____

E _____

Activity 17.2

Lenz's Law

Changing magnetic fields create electric fields. These electric fields can drive an electric current in a loop of wire. This is the principle behind the generation of electricity in electrical generators. The direction of the induced current is given by Lenz's law: *The direction of induced current is such that it opposes the change that produces it.* In this activity you will gain experience using Lenz's law. Review chapter 17 as necessary.

TRUE OR FALSE

_____ Holding a bar magnet near a coil of wire will make current flow through the wire.

_____ Only increasing magnetic fields generate electric fields.

_____ Lenz's law is actually a statement of conservation of energy.

_____ Tilting a loop while it is in a magnetic field will generate current in that loop.

Consider the primary coil of wire connected to a battery as shown. The battery drives an electrical current and the direction is indicated by an arrow. We are interested in the direction of the current induced in the secondary loop on the right. Indicate + or − for the direction of the induced current in the following situations.

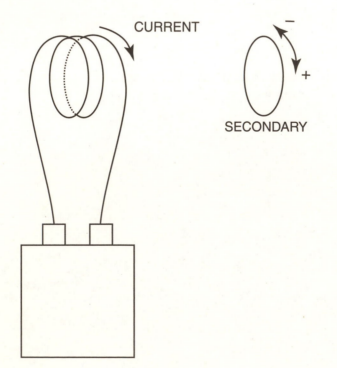

_____ The primary coil is moved towards the secondary coil.

_____ The secondary coil moves towards the primary coil.

_____ The primary coil is turned around 180 degrees.

_____ The primary coil is shut off.

Consider dropping a bar magnet through a circular loop of wire, as shown in the figure.

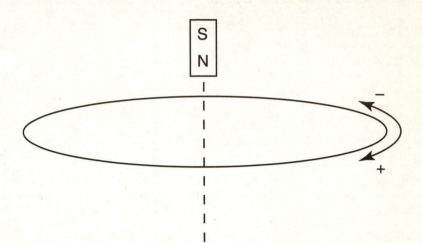

Which graph below would represent the current as a function of time through the circular wire? We take positive current as clockwise when looking down on the loop. _____

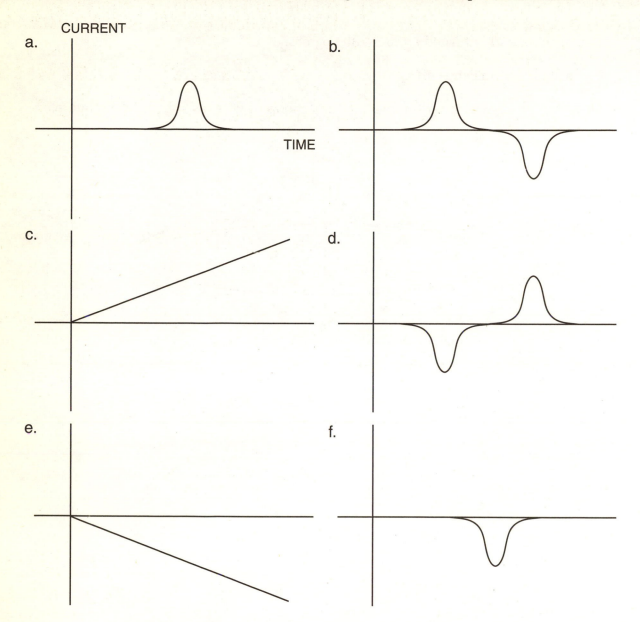

Activity 18.1

Electric Circuits

Electric circuits can be found literally everywhere in your daily life. What are the three essential parts of an electric circuit?

_____

_____

If an electric circuit is carrying a current of 1/2 ampere, how many electrons pass a given point in that circuit in a time of 2 seconds? _____

What is the unit of electrical resistance? _____

The figure shows a magnified view of three pieces of copper wire. If a battery is connected to the ends of the wire, electric current will flow through the wire.

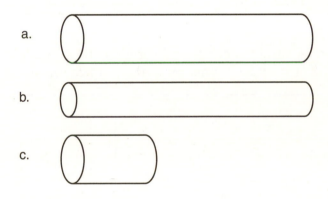

a.

b.

c.

In which wire will the current be the greatest? Explain.

_____

_____

_____

In which wire will the current be the least? Explain.

_____

_____

_____

A 12 volt battery is connected across a resistor whose resistance can be varied. Which of the following graphs represents the current as a function of resistance for this setup?

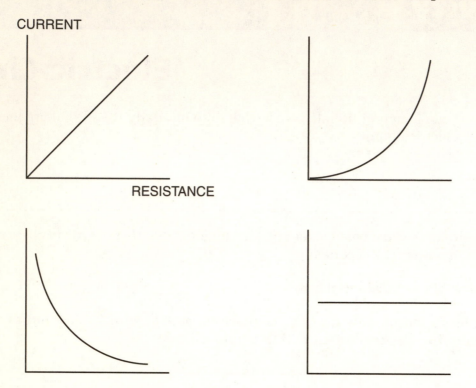

A battery whose voltage can be varied is connected across a 100 ohm resistor. Which of the following graphs represents the current as a function of the voltage for this setup?

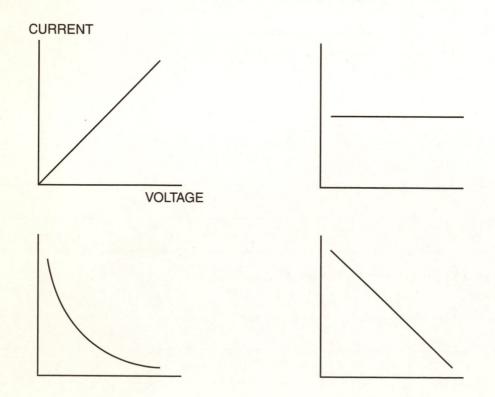

Consider the two circuits pictured below which are lighting identical light bulbs. In fact, the circuits themselves are identical except the one on the right has two 1.5 volt batteries in series, and the one on the left has a single 1.5 volt battery.

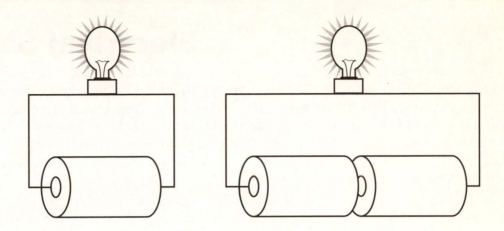

Which light bulb is consuming more power? _____

How much more power is it consuming? Explain.

_____

_____

_____

Activity 18.2

Electrical Safety

Each year in the United States, about 500 people are electrocuted in their homes or workplaces. When in your home you are literally surrounded by potentially dangerous high-voltage sources of electricity. In this activity you will research some aspects of electrical home safety and report your findings. An internet search on "electrical safety" should yield many pages of useful information.

What is a circuit breaker?

_____

_____

_____

_____

What is a fuse?

_____

_____

_____

_____

If an appliance repeatedly blows a fuse or trips a circuit breaker, you should:

a. replace the fuse or circuit breaker with one with a larger current rating.

b. keep replacing the current fuse or resetting the current circuit breaker.

c. plug the appliance into a different outlet.

d. unplug the appliance and have it repaired or replaced.

Why are halogen lamps more of a fire hazard than ordinary lamps?

a. halogen lamps become much hotter than ordinary lamps.

b. halogen lamps use much more current than ordinary lamps.

c. halogen bulbs tend to break, exposing the filament.

d. halogen gas is flammable.

Why is the human body a good conductor of electricity?

_____

_____

_____

What amount of current running through your body would probably be fatal?

_____

How come birds can sit on utility lines safely, but humans are strongly advised not to climb utility poles and touch the lines?

_____

_____

_____

How does a three pronged plug work and why are they used?

_____

_____

_____

What is a polarized plug and why are they used?

_____

_____

_____

What is a "GFCI" and what is its purpose?

_____

_____

_____

How do GFCI's work?

_____

_____

_____

_____

_____

_____

Activity 19.1

Electromagnetic Waves

Electromagnetic waves come in all wavelengths and frequencies. From radio waves to gamma rays, the electromagnetic spectrum is an integral part of our everyday life.

Suppose the three waves below were electromagnetic waves.

a.

b.

c.

Which one has the longest wavelength? _____

Which one has the highest frequency? _____

Which one has the least energy? _____

Which one travels fastest in a vacuum? _____

What is the speed of light? _____

About how long would it take light to travel across the room you are sitting in?

About how long would it take light to travel the distance from New York to California?

How long does it take light to travel from the Sun to the Earth?

The human eye can see only a limited range of electromagnetic waves. What is the range of wavelengths that humans can see? _____

Suppose the human eye could only see in the infrared region of the spectrum. Can you think of any drawbacks to this scenario?

_____

_____

_____

Suppose the human eye was sensitive to the entire spectrum. Can you think of any drawbacks to this scenario?

_____

_____

_____

How come ultraviolet radiation causes sunburn, but infrared does not?

_____

_____

Which of the following statements is true about EM waves?

a. Infrared has more energy than radio waves.

b. X-rays travel faster than visible light.

c. Visible light has a shorter wavelength than ultraviolet.

d. Microwaves have a higher frequency than visible light.

Activity 20.1

Reflection

The figure is an overhead view of a person standing against a wall in a square room. There is a flat mirror on the opposite wall. What parts of the room can he see in the mirror? Indicate these regions by marking the walls of the room.

Consider the cube sitting in front of a flat mirror. If you are looking at the mirror from point P, which face of the cube will you see as you look towards the mirror?

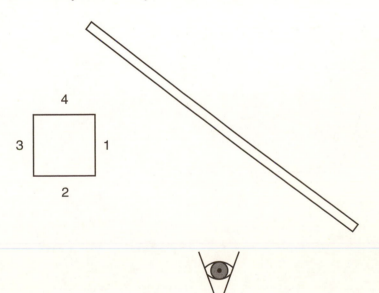

A laser beam is aimed at a convex spherical mirror as shown. Sketch the incident and reflected ray if it is aimed at point A. Do the same for point B.

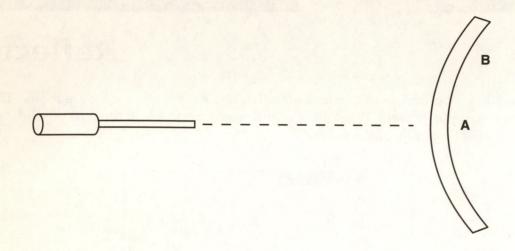

A laser beam is aimed at a concave spherical mirror as shown. Sketch the incident and reflected ray if it is aimed at point A. Do the same for point B.

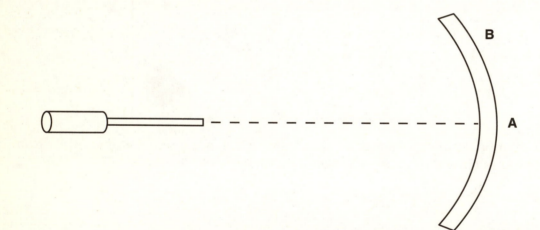

A candle sits near a flat mirror. Where would somebody standing at X see the image of the candle? (Answer D if the image cannot be seen.) _____

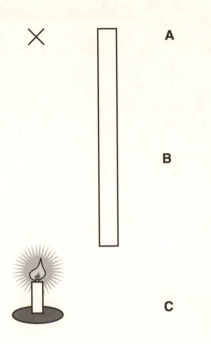

A common Halloween prank is to surprise trick-or-treaters with the 'living head on a table' gag. This can be accomplished with a four legged wooden table and two flat mirrors. How is this trick done? Sketch the setup in the space below.

_____

_____

_____

_____

_____

Activity 20.2

Refraction

When light travels from one medium (water, for example) to another (glass, for example), the light beam changes direction. The amount of direction change depends on the relative indices of refraction of the two media and the incident angle.

1. If the index of refraction of a certain type of glass is 1.5, what is the speed of light in that glass?

2. True or false: When a light beam travels from one medium to another, the light always bends *towards* the normal. _____

3. True or false: Light travels faster in a vacuum than any other material. _____

The figure below depicts a laser beam incident on a piece of glass. The dotted line is the *normal* to the plane of the glass.

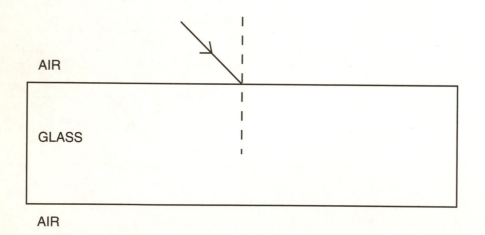

4. Sketch the beam as it travels through the glass and out the other side.

5. If someone were looking up at the glass from point P, where would the laser be seen? Redraw the laser at this location.

6. Is this displacement of the image due to refraction more severe for thick glass or thin glass?

_____

7. Which is faster: running on the beach, or swimming in the water? _____

8. A lifeguard at L has to save a person swimming at X. Draw the path the lifeguard should take so that the she will reach the swimmer in the least amount of time. (If you don't know the answer, do an internet search on "Fermat's principle".)

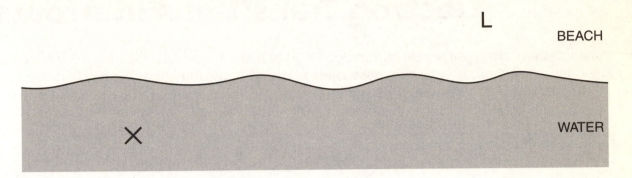

9. What does this lifeguard/swimmer example have to do with light and refraction?

_____

_____

_____

_____

_____

Activity 21.1

Electron Transitions in Atoms

Electrons in atoms occupy discrete energy levels. When an electron jumps from a higher level to a lower level, a photon is emitted whose energy is equal to the energy difference between levels. In the following questions, the energy levels in atoms are represented by horizontal lines, and the distance between the lines is proportional to the energy difference between levels.

1. Suppose an atom has three equally spaced energy levels.

How many possible downward jumps (from higher to lower energy) could an electron make in this atom? _____

How many different photons would result from all possible electron jumps in this atom?

_____

Which jump(s) would result in the highest frequency photon? _____

2. Suppose an atom has four equally spaced energy levels.

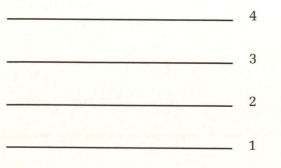

How many possible downward jumps (from higher to lower energy) could an electron make in this atom? _____

How many different photons would result from all possible electron jumps in this atom?

_____

Which jump(s) would result in the longest wavelength photon? _____

3. Suppose an atom has four energy levels, as shown.

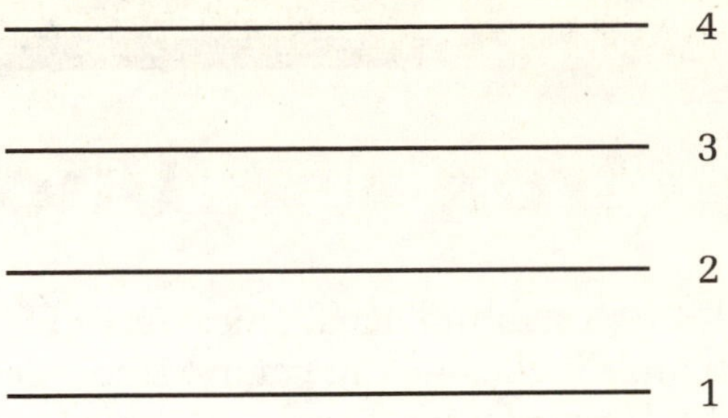

_____ 4

_____ 3

_____ 2

_____ 1

How many possible downward jumps (from higher to lower energy) could an electron make in this atom? _____

How many different photons would result from all possible electron jumps in this atom?

_____

Which jump(s) would result in the highest energy photon? _____

4. Suppose a gas discharge tube emits light that consists of five different wavelength photons. Construct an energy level diagram (as shown above) that would be consistent with an atom that emits five different photons.

5. Repeat number 4 for a discharge tube that emits six different photons.

Activity 22.1

Uncertain Times

One of the central principles of quantum mechanics is Heisenberg's uncertainty principle. Basically the uncertainty principle says that you cannot simultaneously determine the position and momentum of an object with perfect accuracy. Because of the extremely small value of Planck's constant, the uncertainty principle only manifests itself for very small objects (like atoms and subatomic particles). In this activity we will see how the uncertainty principle would affect everyday life by assuming Planck's constant is **much** larger: $h = 1\text{kg} \cdot \text{m}^2/\text{s}$ (instead of the correct value of $h = 6.63 \times 10^{-34}\text{kg} \cdot \text{m}^2/\text{s}$). The uncertainty principle reads, $\Delta x \times p > h$. Since momentum (p) equals mass times velocity, we can write this as $m \times \Delta x \times \Delta v > h$, or

$$\Delta x \times \Delta v > h/m$$

1. If you took a meter stick and measured the position of a baseball on your desktop to an accuracy of $\pm 1\text{mm}$ or $\pm 0.001\text{mm}$, what is the uncertainty in the baseball's velocity? (Assume the baseball has a mass of 0.5kg.)

Ordinarily you could just look at the baseball and be certain that its velocity is zero. What does the above result mean?

_____

_____

_____

_____

2. As you drive down the road, you look at your speedometer and you see that you are going 10 miles an hour. The uncertainty in your knowledge of your speed is ±0.5 miles/hour, which is about ±0.25 m/s. What is the uncertainty in the position of your car? (Assume the car has a mass of 1000 kg.)

Does this seem reasonable? Do you think it's possible, with modern day equipment, to measure the position of a car to greater accuracy?

_____

_____

_____

_____

_____

3. You see a large wasp in your bedroom and sprint out and close the door. Your bedroom measures 5 meters on a side. As you stand outside your bedroom, wondering what to do, what is the uncertainty in the wasp's speed? (Assume the wasp has a mass of 5 grams.)

Does it make sense that you would not know the speed of the wasp if you're not in the room with it? Does this mean that the uncertainty principle actually does apply to large objects? Explain.

_____

_____

_____

_____

4. Make up your own example of an application of the uncertainty principle to everyday objects. What is the situation?

_____

_____

_____

_____

What is the uncertainty?

Discuss your result.

_____

_____

_____

_____

Activity 22.2

Probability Graphs

The *wave function* of a quantum object is a set of probabilities that predicts the likelihood of finding the object in a particular location. Wave functions can be represented on probability graphs, as discussed in chapter 22. In the following exercises, you will be given a description of an object and its situation. For each object, make an approximate probability graph. The position of the object will be denoted by x. Be sure to label your x axis.

1. A marble is confined to the interior of a rounded bowl. If this is all the information you have available, construct an approximate probability graph for the location of the marble.

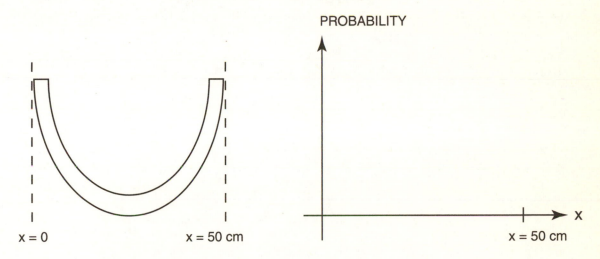

2. A metal cube 10 cm on a side is completely evacuated except for a single hydrogen atom. You have no information about the whereabouts of the atom except that is it inside the box.

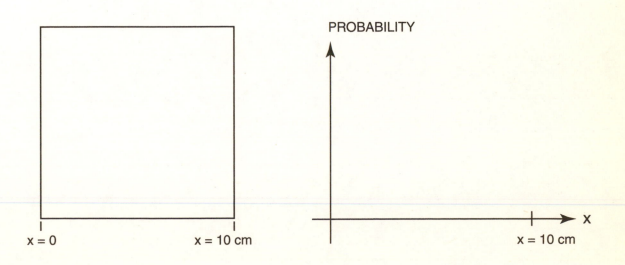

3. John lives alone in a house in the woods. Except for occasional trips to the outhouse, he spends all of his time inside the house.

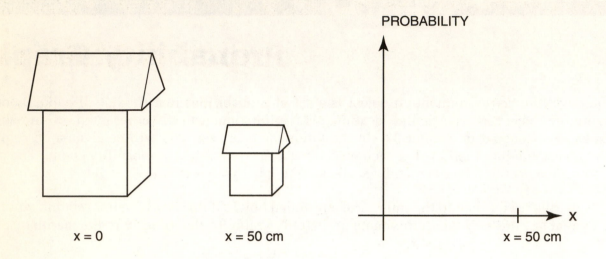

4. Make up your own example in the space below.

Activity 23.1

Composite Materials

Composite materials combine the properties of two or more materials. Usually the purpose is to create a material that has more strength. In this activity you will research some composite materials and report your findings.

COMPOSITE CORE POWER LINES

Describe the construction of traditional power lines.

_____

_____

_____

What is a composite core power line? What materials are used?

_____

_____

_____

_____

What is the primary advantage of composite core power lines over traditional power lines? Can they carry more electricity?

_____

_____

_____

_____

FIBERGLASS

What are the raw materials used in making fiberglass?

_____

_____

Find at least two examples of construction materials that fiberglass has largely replaced in the last 30 years.

_____

_____

What is the advantage of fiberglass over the other materials?

_____

_____

_____

SAFETY GLASS

Safety glass is frequently used in automobile applications. What is safety glass?

_____

_____

_____

How is safety glass constructed?

_____

_____

_____

_____

BULLETPROOF GLASS

What is the difference between bulletproof glass and safety glass?

_____

_____

_____

How is bulletproof glass constructed?

_____

_____

_____

_____

Activity 24.1

Magnetic Recording

One of the most important applications of ferromagnetic materials is in magnetic recording devices like computer disks and magnetic recording tape. In this activity you will be asked to research magnetic recording and report on your findings.

1. What is a ferromagnetic material?

   _____

   _____

   _____

   _____

2. What is a hysteresis curve?

   _____

   _____

3. Sketch a hysteresis curve on the graph below.

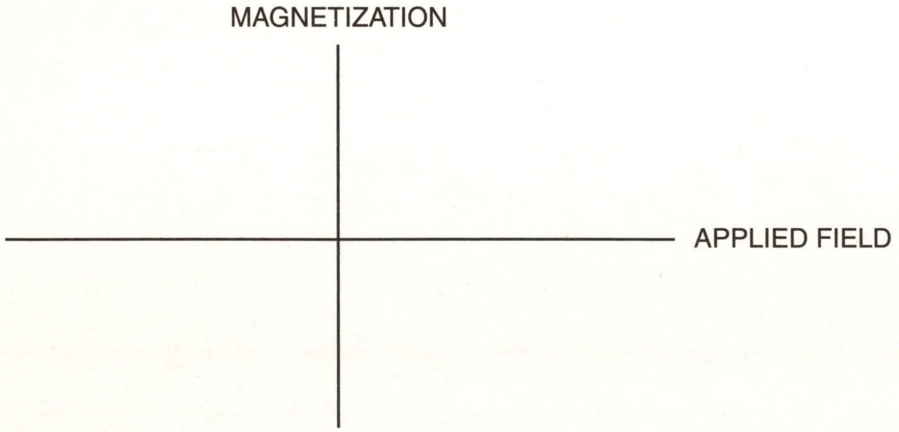

MAGNETIZATION

APPLIED FIELD

4. Explain the meaning of the hysteresis curve.

_____

_____

_____

_____

_____

_____

_____

5. What is *remnant magnetization*?

_____

_____

_____

6. Describe the role of the electromagnet in the magnetic recording system recording system.

_____

_____

_____

_____

_____

7. Are writeable CD's an example of magnetic recording? Explain.

_____

_____

_____

_____

Activity 25.1

Binary

Computers store information in binary form. All letters and numbers can be represented by a series of 1's and 0's. In chapter 25 we learned that the 26 letters of the alphabet can be represented by 5 bits of information. Create your own binary representation of the alphabet by filling in the table below.

LETTER	BINARY CODE	LETTER	BINARY CODE
A		N	
B		O	
C		P	
D		Q	
E		R	
F		S	
G		T	
H		U	
I		V	
J		W	
K		X	
L		Y	
M		Z	

Numbers can be represented in binary also. The number 13 is represented by 1011. Each digit represents a power of 2:

$$1 \quad 1 \quad 0 \quad 1$$

$$1 \times 2^3 + 1 \times 2^2 + 0 \times 2^1 + 1 \times 2^0 = 13$$

To help with the rest of this activity, calculate the following numbers with your calculator:

$2^0 =$ $\qquad$ $2^5 =$ $\qquad$ $2^{10} =$

$2^1 =$ $\qquad$ $2^6 =$ $\qquad$ $2^{11} =$

$2^2 =$ $\qquad$ $2^7 =$ $\qquad$ $2^{12} =$

$2^3 =$ $\qquad$ $2^8 =$

$2^4 =$ $\qquad$ $2^9 =$

Fill in the following table of numbers with their binary equivalents.

NUMBER	BINARY EQUIVALENT
0	
10	
18	
43	
56	
544	
1026	
2450	
1111	
5325	

Activity 26.1

Nuclear Dimensions

In chapter 26 you learned that atoms are mostly empty space. The atom consists of a nucleus and electrons orbiting it. Consider the schematic diagram of an atom below. In this activity we will consider the size of your body if we eliminated all of the empty space in it by packing the nuclei close together.

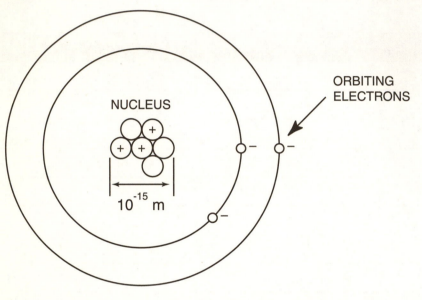

1. Assuming the atom is spherical, calculate the volume of the atom. (The volume of a sphere is $\frac{4}{3}\pi R^3$, where R is the radius of the sphere.)

2. Assuming the nucleus is spherical, calculate the volume of the nucleus.

3. If this atom is representative of atoms found in normal matter, what fraction of normal matter is empty space? What fraction is actually matter?

4. Estimate the volume of your body in cubic meters. (Assume you are rectangular and estimate your height, depth, and width.)

5. Estimate the volume of actual matter in your body using the answer to number 3. If all of the matter (the atomic nuclei) in your body were packed into a small sphere, what would the radius of this sphere be?

6. Could all of the matter in your body fit onto the head of a pin? Explain.

   _____

   _____

   _____

   _____

   _____

7. What would be the practical difficulties in trying to pack nuclei very close together?

   _____

   _____

   _____

   _____

Activity 26.2

Radioactivity

How many protons and neutrons are in the following nuclei?

NUCLEUS	# PROTONS	# NEUTRONS
Carbon-12		
Silicon-24		
Iron-59		
Nickel-55		
Promethium-138		
Ytterbium-160		

When a radioactive nucleus decays by alpha or beta decay, it changes into a new element. Determine the product of the following radioactive decays.

RADIOACTIVE NUCLEUS	DECAY MODE	DAUGHTER NUCLEUS
Hydrogen-3	beta	
Carbon-14	beta	
Oxygen-15	beta	
Hafnium-156	alpha	
Radon-212	alpha	
Thorium-228	alpha	
Americium-241	alpha	
Americium-244	beta	
Uranium-238	alpha	
Sodium-25	beta	

The half-life of a radioactive nucleus is the amount of time it takes for half of a large sample of those nuclei to decay.

CARBON-14

Carbon-14 decays by beta decay. What does carbon-14 turn into when it decays?

The half-life of carbon-14 is roughly 5000 years. Assuming you start with a 10 gram sample of carbon-14, plot the amount left versus time on the graph below.

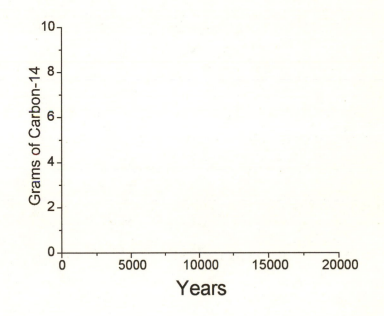

HYDROGEN-3

Hydrogen-3 (also known as tritium) decays by beta decay. What does hydrogen-3 turn into when it decays?

The half-life of hydrogen-3 is roughly 12 years. Assuming you start with a 64 gram sample of hydrogen-3, plot the amount left versus time on the graph below.

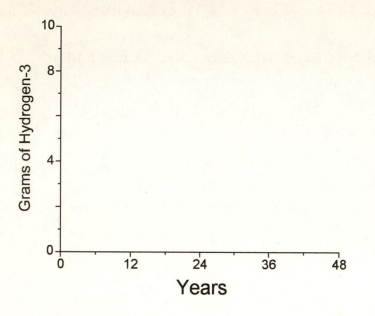

Activity 26.3

Radioactivity Around The Home

AMERICIUM-241

Most smoke detectors contain an artificially produced isotope: americium-241. Do some research into the operation of smoke detectors and answer the following questions.

1. Why do smoke detectors use radioactive isotopes? Describe the function of radioactivity in the operation of the smoke detector.

_____

_____

_____

_____

_____

_____

_____

_____

_____

_____

2. How is americium formed?

_____

_____

_____

_____

_____

3. What is the half-life of americium-241? Based on this half-life, how long would you expect the smoke detector to operate without having to replace the radioactive source inside?

_____

_____

_____

4. If possible, find the manual for a smoke detector. Photocopy it and staple it to this page before turning it in.

5. Do you think the additional safety of using smoke detectors outweighs the risks associated with having americium-241 in the home? Explain.

_____

_____

_____

RADON

Radon-222 is a radioactive gas that is associated with certain types of cancer. Many homes around the world have levels of radon that are considered unsafe.

6. Where does radon enter the home?

_____

_____

7. Why is radon considered more dangerous to smokers than non-smokers?

_____

_____

_____

_____

8. How is radon-222 created? (What parent nucleus decays into radon-222?)

_____

_____

9. How does a radon testing device work?

_____

_____

_____

_____

_____

10. If unsafe levels of radon are found in your home, what can be done?

_____

_____

_____

_____

_____

Activity 26.4

Fission and Critical Masses

A *critical mass* is the correct mass and shape of fissionable material that will sustain a chain reaction. A 52 kilogram sphere of uranium-235 is barely a critical mass. That is, if it were slightly smaller it would not support a self-sustaining chain reaction. Investigate the concept of critical mass on the internet. Explain your findings below.

Consider the following configurations of uranium-235. Would you expect any of these to be a critical mass? Explain your reasoning.

52 kilogram disk of uranium-235:

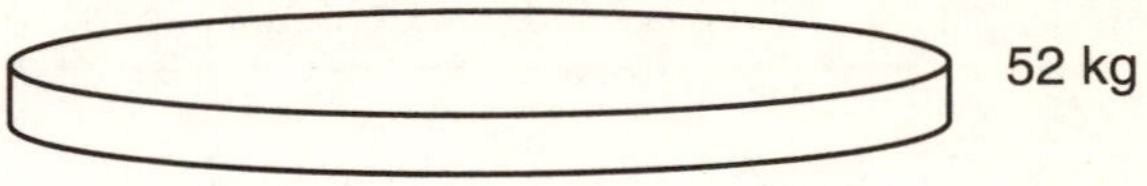

52 kg

_____

_____

_____

_____

52 kilogram cube of uranium-235:

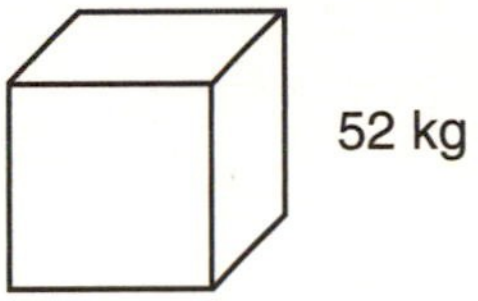

52 kg

_____

_____

_____

_____

60 kilogram sphere of uranium-235:

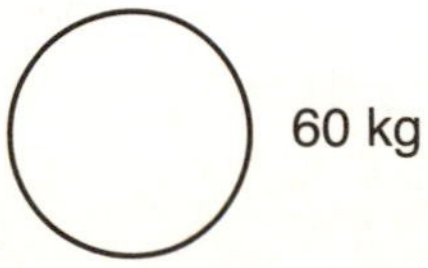

60 kg

_____

_____

_____

_____

50 kilogram sphere of uranium-235:

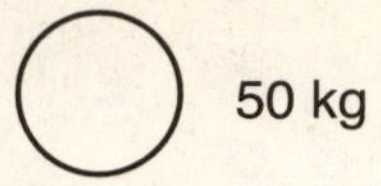

◯ 50 kg

_____

_____

_____

_____

_____

Activity 27.1

Particle Tracks

The goal of high energy particle physics is to discover the types of elementary particles and how they interact with each other. In a typical experiment, particles (protons or electrons, for example) are given tremendous kinetic energy by a particle accelerator. Then they are forced to collide with each other, or with another target producing a number of other particles in the process. One of the most complex aspects of this type of experiment is the detection of the particles created in the collisions. The tracks they leave in the detector tell physicists much about the collision.

The following figures represent tracks of particles in a region where a magnetic field exists. The magnetic field points into the page.

1. Particle A, moving left to right, decays into particles B and C.

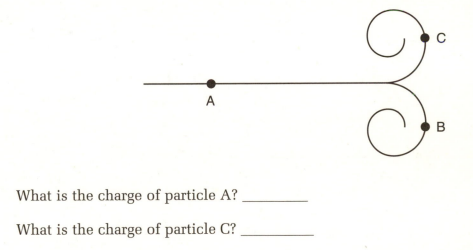

What is the charge of particle A? _____

What is the charge of particle C? _____

2. Particle A, moving left to right, decays into particles B and C.

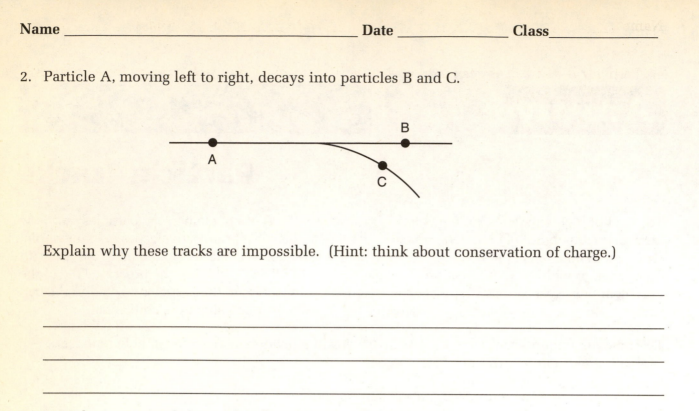

Explain why these tracks are impossible. (Hint: think about conservation of charge.)

_____

_____

_____

_____

3. Particle A, moving left to right, decays into particles B and C.

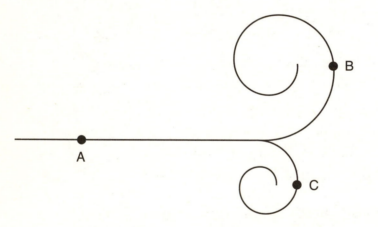

Assuming B and C have equal and opposite charges, which one is more massive? Explain.

_____

_____

_____

4. Particles A and B come together horizontally, annihilate, and two photons leave vertically, as shown.

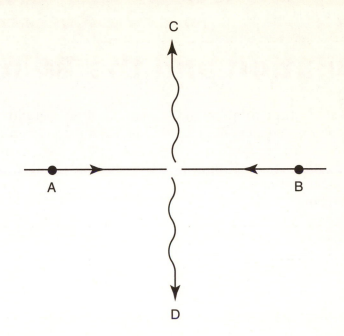

Compare the charge of A to the charge of B.

_____

_____

Can you say anything else about particles A and B?

_____

_____

Compare the wavelength of photon C to photon D. Explain.

_____

_____

_____

Activity 28.1

Time Dilation and the Twin Paradox

In this activity you will study time dilation and the famous Twin Paradox in Special Relativity. Be sure to review the *Looking Deeper: The Size of Time Dilation* section in the text.

Imagine that your twin brother takes off in a spaceship and travels at 99.5% the speed of light. According to your ground based clock, he is gone for one year. How much time (according to you) passes on the spaceship?

Now imagine what your brother sees. In his reference frame he sees himself as stationary and he sees you (the Earth) moving at 99.5% the speed of light. According to his watch, how many Earth years pass for each spaceship year?

Why is this situation called a paradox?

_____

_____

_____

_____

_____

Do some research on the internet and find out how this paradox is resolved. Write your findings in the space below.

_____

_____

_____

_____

_____

_____

_____

_____

_____

What is the key difference between special relativity and general relativity?

_____

_____

_____

Activity 28.2

A Relativity Quiz

1. You've decided to let your sister, a NASA astronaut, cook the thanksgiving turkey this year. Normally, the turkey will take 6 hours to cook, but your sister decides to cook it on her spaceship while traveling at close to the speed of light. According to your watch, she was gone for 6 hours. The turkey comes back:

   a. overcooked

   b. undercooked

   c. just right

   d. both undercooked and overcooked

2. According to a person on the surface of the Earth, the clock in a GPS satellite runs _____ because of its relative motion, and _____ because of its higher altitude.

   a. slow, slow

   b. slow, fast

   c. fast, slow

   d. fast, fast

3. Because of relative motion, you notice a friend's clock running slowly. Your friend notices that your clock is running:

   a. fast

   b. slowly

   c. normal

4. Someone shines a light while moving towards you at 1000 m/s. With what speed will the light strike you? (The speed of light is 300,000,000m/s.)

   a. 300,000,000m/s

   b. 300,001,000m/s

   c. 299,999,000m/s

   d. any of these are possible

5. If you are in a spaceship far from the reaches of gravity, under what conditions will you feel weightless?

   a. if it is accelerating at just the right rate

   b. if it is moving uniformly

   c. if is spiraling out of control

6. You are riding on a flatbed truck moving at 50 miles per hour. You have two guns, one aimed forward and one aimed backwards and you fire them at the same time. What is the fastest observed speed?

   a. forward moving bullet in the truck's reference frame

   b. forward moving bullet in the ground's reference frame

   c. backward moving bullet in the truck's reference frame

   d. backward moving bullet in the ground's reference frame

7. If you can do 20 pushups on the surface of the Earth, how many can you do in a spaceship, far from gravity, accelerating at g (10 m/s/s)?

   a. 20

   b. less than 20

   c. more than 20

   d. probably hundreds

8. You are jealous of your younger brother, who looks very young for his age. You are interested in reducing the rate at which you age, relative to him. Given the choice, would you work as a park ranger, high in the mountains, or as a taxi cab driver at sea level? (Take into account the effects of gravity only.) Explain your reasoning.

   _____

   _____

   _____

   _____

9. It is well-known that light bends the same amount per second as a thrown baseball. How come a flashlight beam appears to travel in a straight line path?

_____

_____

_____

Activity 29.1

Astronomical Distances and Hubble's Law

The distances we measure in everyday life range from millimeters to kilometers. However, to measure astronomical distances it is more convenient to measure distance in light years of parsecs. Estimate, in years, how long it would take for you to walk one light year. Show your calculation.

Edwin Hubble found evidence that the universe was expanding. (Review Hubble's Law and the redshift in chapter 29.) Suppose that his data were summarized in the following table (instead of Table 29-1):

DISTANCE TO GALAXY (MPC)	VELOCITY (KM/S)
1	250
1.5	750
2	1000
2.5	1300
3	1700

Plot this data in the graph below.

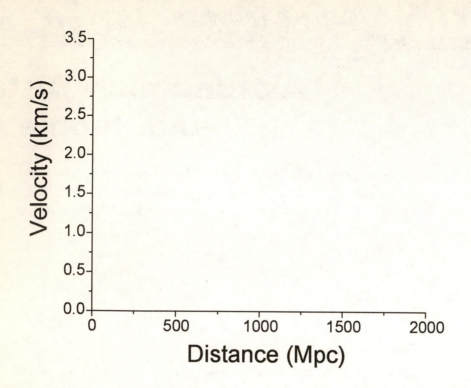

Draw a straight line that best represents this data and calculate the Hubble constant.

What is the approximate age of the universe based on this data?